CAMBRIDGE LIBRARY COLLECTION

Books of enduring scholarly value

Darwin

Two hundred years after his birth and 150 years after the publication of 'On the Origin of Species', Charles Darwin and his theories are still the focus of worldwide attention. This series offers not only works by Darwin, but also the writings of his mentors in Cambridge and elsewhere, and a survey of the impassioned scientific, philosophical and theological debates sparked by his 'dangerous idea'.

On the Movements and Habits of Climbing Plants

Initially published by the Linnean Society, this 1865 essay was Darwin's first foray into the study of climbing plants. He was inspired to produce this work by a paper on the tendrilled Cucurbitacean plant by American botanist Asa Gray, with whom he had a firm intellectual friendship. Darwin examines in detail those plants which climb using a twisting stem, such as the hop; leaf-climbers, such as the clematis; tendrilled plants such as the passion flower; and hook and root climbers such as ivy. The conclusions reached by his study are presented in terms of the adaptations of various species to their environments, a continuation of the theories that Darwin had propounded in his Origin of the Species six years earlier. His passion for the design of the plants and fascination with the diversity of their powers of movement are clear in this accessible example of the process of evolution.

Cambridge University Press has long been a pioneer in the reissuing of out-of-print titles from its own backlist, producing digital reprints of books that are still sought after by scholars and students but could not be reprinted economically using traditional technology. The Cambridge Library Collection extends this activity to a wider range of books which are still of importance to researchers and professionals, either for the source material they contain, or as landmarks in the history of their academic discipline.

Drawing from the world-renowned collections in the Cambridge University Library, and guided by the advice of experts in each subject area, Cambridge University Press is using state-of-the-art scanning machines in its own Printing House to capture the content of each book selected for inclusion. The files are processed to give a consistently clear, crisp image, and the books finished to the high quality standard for which the Press is recognised around the world. The latest print-on-demand technology ensures that the books will remain available indefinitely, and that orders for single or multiple copies can quickly be supplied.

The Cambridge Library Collection will bring back to life books of enduring scholarly value across a wide range of disciplines in the humanities and social sciences and in science and technology.

On the Movements and Habits of Climbing Plants

CHARLES DARWIN

CAMBRIDGE
UNIVERSITY PRESS

CAMBRIDGE UNIVERSITY PRESS

Cambridge New York Melbourne Madrid Cape Town Singapore São Paolo Delhi

Published in the United States of America by Cambridge University Press, New York

www.cambridge.org
Information on this title: www.cambridge.org/9781108003599

This edition first published 1865
This digitally printed version 2009

ISBN 978-1-108-00359-9

ON THE

MOVEMENTS AND HABITS

OF

CLIMBING PLANTS.

BY

CHARLES DARWIN, Esq., F.R.S., F.L.S. etc.

[*From the* JOURNAL OF THE LINNEAN SOCIETY.]

LONDON:
SOLD AT THE SOCIETY'S APARTMENTS, BURLINGTON HOUSE;
AND BY
LONGMAN, GREEN, LONGMAN, ROBERTS & GREEN,
AND
WILLIAMS AND NORGATE.
1865.

ON THE

MOVEMENTS AND HABITS

OF

CLIMBING PLANTS.

————————

TABLE OF CONTENTS.

	PAGE
Introduction	1

Part I.—SPIRALLY TWINING PLANTS.

	PAGE
Axial twisting	5
Nature of the revolving movement	7
Purpose of the revolving movement, and manner of the spiral ascent	9
Table of the rates of revolution	14
Anomalous revolvers	21
Variations in the power of twining	24

Part II.—LEAF-CLIMBERS.

	PAGE
Clematis	26
Tropæolum	34
Antirrhineæ	38
Solanum	41
Fumariaceæ	43
Cocculus	45
Gloriosa	45
Flagellaria	46
Nepenthes	46
Summary on Leaf-climbers	47

Part III.—TENDRIL-BEARERS.

	PAGE
Bignoniaceæ	49
Polemoniaceæ	61
Leguminosæ	65
Compositæ	67
Smilaceæ	68
Fumariaceæ	70
Cucurbitaceæ	73
Vitaceæ	79
Sapindaceæ	87
Passifloraceæ	89
Spiral contraction of tendrils	92
Summary of the nature and action of tendrils	98

Part IV.—HOOK- AND ROOT-CLIMBERS; CONCLUDING REMARKS.

	PAGE
Hook-climbers	105
Root-climbers	105
Concluding remarks on Climbing-plants	107

I WAS led to this subject by an interesting, but too short, paper by Professor Asa Gray on the movements of the tendrils of some Cucurbitaceous plants*. My observations were more than half completed before I became aware that the surprising phenomenon of the spontaneous revolutions of the stems and tendrils of climbing plants had been long ago observed by Palm and by Hugo von Mohl†, and had subsequently been the subject of two

* Proc. Amer. Acad. of Arts and Sciences, vol. iv. Aug. 12, 1858, p. 98.

† Ludwig H. Palm, Ueber das Winden der Pflanzen; Hugo von Mohl, Ueber den Bau und das Winden der Ranken und Schlingpflanzen, 1827. Palm's

memoirs by Dutrochet*. Nevertheless I believe that my observations, founded on the close examination of above a hundred
widely distinct living plants, contain sufficient novelty to justify
me in laying them before the Society.

Climbing plants may be conveniently divided into those which
spirally twine round a support, those which ascend by the movement of the foot-stalks or tips of their leaves, and those which
ascend by true tendrils,—these tendrils being either modified
leaves or flower-peduncles, or perhaps branches. But these subdivisions, as we shall see, nearly all graduate into each other.
There are two other distinct classes of climbing-plants, namely
those furnished with hooks and those with rootlets; but, as such
plants exhibit no special movements, we are but little concerned
with them; and generally, when I speak of climbing plants, I refer
exclusively to the first great class.

Part I.—Spirally twining Plants.

This is the largest subdivision, and is apparently the primordial and simplest condition of the class. My observations will be
best given by taking a few special cases. When the shoot of a
Hop (*Humulus Lupulus*) rises from the ground, the two or three
first-formed internodes are straight and remain stationary; but
the next-formed, whilst very young, may be seen to bend to one
side and to travel slowly round towards all points of the compass,
moving, like the hands of a watch, with the sun. The movement
very soon acquires its full ordinary velocity. From seven observations made during August on shoots proceeding from a plant
which had been cut down, and on another plant during April, the
average rate during hot weather and during the day was 2 h. 8 m.
for each revolution; and none of the revolutions varied much
from this rate. The revolving movement continues as long as
the plant continues to grow; but each separate internode, as it
grows old, ceases to move.

To ascertain more precisely what amount of movement each internode underwent, I kept a potted plant in a well-warmed room
to which I was confined during the night and day. A long inclined shoot projected beyond the upper end of the supporting

Treatise was published only a few weeks before Mohl's. See also 'The Vegetable Cell' (translated by Henfrey), by H. von Mohl, p. 147 to end.
   * "Des Mouvements révolutifs spontanés," &c., 'Comptes Rendus,' tom. xvii.
(1843) p. 989; "Recherches sur la Volubilité des Tiges," &c., tom. xix. (1844)
p. 295.

stick, and was steadily revolving. I then took a longer stick and tied up the shoot, so that only a very young internode, $1\frac{3}{4}$ of an inch in length, was left free; this was so nearly upright that its revolution could not be easily observed; but it certainly moved, and the side of the internode which was at one time convex became concave, which, as we shall hereafter see, is a sure sign of the revolving movement. I will assume that it made at least one revolution during the first twenty-four hours. Early the next morning its position was marked, and it made the second revolution in 9 h.; during the latter part of this revolution it moved much quicker, and the third circle was performed in the evening in a little over 3 h. As on the succeeding morning I found that the shoot revolved in 2 h. 45 m., it must have made during the night four revolutions, each at the average rate of a little over 3 h. I should add that the temperature of the room varied only a little. The shoot had now grown $3\frac{1}{2}$ inches in length, and carried at its extremity a young internode 1 inch in length, which showed slight changes in its curvature. The next or ninth revolution was effected in 2 h. 30 m. From this time forward, the revolutions were easily observed. The thirty-sixth revolution was performed at the usual rate; so was the last or thirty-seventh, but it was not quite completed; for the internode abruptly became upright, and, after moving to the centre, remained motionless. I tied a weight to its upper end, so as to slightly bow it, and thus to detect any movement; but there was none. Some time before the last revolution the lower part of the internode had ceased to move.

A few more remarks will complete all that need be said on this one internode. It moved during five days; but the more rapid movement after the third revolution lasted during three days and twenty hours. The regular revolutions, from the ninth to thirty-sixth inclusive, were performed at the average rate of 2 h. 31 m.: the weather was cold; and this affected the temperature of the room, especially during the night, and consequently retarded a little the rate of movement. There was only one irregular movement, when a segment of a circle was rapidly performed (not counted in the above enumeration); and this occurred after an unusually slow revolution of 2 h. 49 m. After the seventeenth revolution the internode had grown from $1\frac{3}{4}$ to 6 inches in length, and carried an internode $1\frac{7}{8}$ inch long, which was just perceptibly moving; and this carried a very minute ultimate internode. After the twenty-first revolution, the penultimate internode was $2\frac{1}{2}$ inches long, and probably revolved in a period of about three hours. At the twenty-seventh revolution our lower internode was $8\frac{3}{8}$, the penultimate $3\frac{1}{2}$, and ultimate $2\frac{1}{2}$ inches in length; and the inclination of the whole shoot was such, that a circle 19 inches in diameter was swept by it. When the movement ceased, the lower internode was 9 and the penultimate 6 inches in length; so that, from the twenty-seventh to thirty-seventh revolutions inclusive, three internodes were at the same time revolving.

The lower internode, when it ceased revolving, became upright and rigid; but as the whole shoot continued to grow unsupported, it became nearly horizontal, the uppermost and growing internodes still revolving at the extremity, but of course no longer round the old central point of the supporting stick. From the change in the position of the centre of gravity of the revolving extremity, a slight and slow swaying movement was given to the long and horizontally projecting shoot, which I mistook at first for a spontaneous movement. As the shoot grew, it depended more and more, whilst the growing and revolving extremity turned itself up more and more.

With the Hop we have seen that three internodes were at the same time revolving; and this was the case with most of the plants observed by me. With all, if in full health, two revolved; so that by the time one had ceased, that above it was in full action, with a terminal internode just commencing to revolve. With *Hoya carnosa*, on the other hand, a depending shoot, 32 inches in length, without any developed leaves, and consisting of seven internodes (a minute terminal one, an inch in length, being counted), continually, but slowly, swayed from side to side in a semicircular course, with the extreme internodes making complete revolutions. This swaying movement was certainly due to the movement of the lower internodes, which, however, had not force sufficient to swing the whole shoot round the central supporting stick. The case of another Asclepiadaceous plant, viz. *Ceropegia Gardnerii* is worth briefly giving. I allowed the top to grow out almost horizontally to the length of 31 inches; this now consisted of three long internodes, terminated by two short ones. The whole revolved in a course opposed to the sun (the reverse of that of the Hop), at rates between 5 h. 15 m. and 6 h. 45 m. for each revolution. Hence, as the extreme tip made a circle of above 5 feet (or 62 inches) in diameter and 16 feet in circumference, the tip travelled at the rate (assuming the circuit to have been completed in six hours) of 32 or 33 inches per hour. The weather being hot, the plant was allowed to stand on my study-table; and it was an interesting spectacle to watch the long shoot sweeping, night and day, this grand circle in search of some object round which to twine. If we take hold of a growing sapling, we can of course bend it so as to make its tip describe a circle, like that performed by the tip of a spontaneously revolving plant. By this movement the sapling is not in the least twisted round its own axis. I mention this because if a black point be painted on the bark, on the side which is uppermost when the sapling is bent towards the holder's body, as the circle is described, the black point gradually turns round and 'sinks to the lower side, and comes up again when the circle is completed; and this gives the false appearance of twisting, which, in the case of spontaneously revolving plants, deceived me for a time. The appearance is the more deceitful because the axes of nearly all twining-plants are really twisted; and they are twisted in the same direction with the spontaneous revolving movement. To give an instance, the internode of the Hop of which the history has been recorded was at first, as could be 'seen by the ridges on its surface, not in the least twisted; but when, after the 37th revolution, it had grown 9 inches long, and its revolving movement had ceased, it had become twisted three times round its own axis, in the line of the course of the sun; on the other hand, the common Convolvulus, which revolves in an opposite course to the Hop, becomes twisted in an opposite direction.

Hence it is not surprising that Hugo von Mohl (S. 105, 108, &c.) thought that the twisting of the axis caused the revolving movement. I cannot fully understand how the one movement is supposed to cause the other; but it is scarcely possible that the twisting of the axis of the Hop three times could have caused thirty-seven revolutions. Moreover, the revolving movement commenced in the young internode before any twisting of the axis could be detected; and the internode of a young Siphomeris or Lecontea revolved during several days, and became twisted only once on its own axis. But the best evidence that the twisting does not cause the revolving movement is afforded by many leaf-climbing and tendril-bearing plants (as *Pisum sativum, Echinocystis lobata, Bignonia capreolata, Eccremocarpus scaber*, and with the leaf-climbers, *Solanum jasminoides* and various species of *Clematis*), of which the internodes are not regularly twisted, but which regularly perform, as we shall hereafter see, revolving movements like those of true twining-plants. Moreover, according to Palm (S. 30, 95) and Mohl (S. 149), and Léon *, internodes may occasionally, and even not very rarely, be found which are

* Bull. Bot. Soc. de France, tom. v. 1858, p. 356.

twisted in an opposite direction to the other internodes on the same plant, and to the course of revolution; and this, according to Léon (p. 356), is the case with all the internodes of a variety of the *Phaseolus multiflorus*. Internodes which have become twisted round their own axes, if they have not ceased revolving, are still capable of twining, as I have several times observed.

Mohl has remarked (S. 111) that when a stem twines round a smooth cylindrical stick, it does not become twisted. Accordingly I allowed kidney-beans to run up stretched string, and up smooth rods of iron and glass, one-third of an inch in diameter, and they became twisted only in that degree which follows as a mechanical necessity from the spiral winding. The stems, on the other hand, which had ascended the ordinary rough sticks were all more or less and generally much twisted. The influence of the roughness of the support in causing axial twisting was well seen in the stems which had twined up the glass rods; for these were fixed in split sticks below, and were secured above to cross sticks, and the stems in passing these places became very much twisted. As soon as the stems which had ascended the iron rods reached the summit and became free, they also became twisted; and this apparently occurred more quickly during windy weather. Several other facts could be given, showing that the axial twisting stands in relation to inequalities in the support, and likewise to the shoot revolving freely without any support. Many plants, which are not twiners, become in some degree twisted round their own axes[*]; but this occurs so much more generally and strongly with twining-plants than with other plants, that there must be some connexion between the capacity for twining and axial twisting. The most probable view, as it seems to me, is that the stem twists itself to gain rigidity (on the same principle that a much twisted rope is stiffer than a slackly twisted one), so as to be enabled either to pass over inequalities in its spiral ascent, or to carry its own weight when allowed to revolve freely[†].

---

[*] Professor Asa Gray has remarked to me, in a letter, that in *Thuja occidentalis* the twisting of the bark is very conspicuous. The twist is generally to the right of the observer; but, in noticing about a hundred trunks, four or five were observed to be twisted in an opposite direction.

[†] It is well known that stems of many plants occasionally become spirally twisted in a monstrous manner; and since the reading of this paper, Dr. Maxwell Masters has remarked to me in a letter that "some of these cases, if not all, are dependent upon some obstacle or resistance to their upward growth." This conclusion agrees with, and perhaps explains, the normal axial twisting of twining-plants; but does not preclude the twisting being of service to the plant and giving greater rigidity to the stem.

I have just alluded to the twisting which necessarily follows from the spiral ascent of the stem, namely, one twist for each spire completed. This was well shown by painting straight lines on stems, and then allowing them to twine; but, as I shall have to recur to this subject under Tendrils, it may be here passed over.

I have already compared the revolving movement of a twining plant to that of the tip of a sapling, moved round and round by the hand held some way down the stem; but there is a most important difference. The upper part of the sapling moves as a rigid body, and remains straight; but with twining plants every inch of the revolving shoot has its own separate and independent movement. This is easily proved; for when the lower half or two-thirds of a long revolving shoot is quietly tied to a stick, the upper free part steadily continues revolving: even if the whole shoot, except the terminal tip of an inch or two in length, be tied up, this tip, as I have seen in the case of the Hop, Ceropegia, Convolvulus, &c., goes on revolving, but much more slowly; for the internodes, until they have grown to some little length, always move slowly. If we look to the one, two, or several internodes of a revolving shoot, they will be all seen to be more or less bowed either during the whole or during a large part of each revolution. Now if a coloured streak be painted (this was done with a large number of twining plants) along, we will say, the convex line of surface, this coloured streak will after a time (depending on the rate of revolution) be found to lie along one side of the bow, then along the concave side, then on the opposite side, and, lastly, again on the original convex surface. This clearly proves that the internodes, during the revolving movement, become bowed in every direction. The movement is, in fact, a continuous self-bowing of the whole shoot, successively directed to all points of the compass.

As this movement is rather difficult to understand, it will be well to give an illustration. Let us take the tip of a sapling and bend it to the south, and paint a black line on the convex surface; then let the sapling spring up and bend it to the east, the black line will then be seen on the lateral face (fronting the north) of the shoot; bend it to the north, the black line will be on the concave surface; bend it to the west, the line will be on the southern lateral face; and when again bent to the south, the line will again be on the original convex surface. Now, instead of bending the sapling, let us suppose that the cells on its whole southern surface were to contract from the base to the tip, the whole shoot would be bowed to the south; and let the longitudinal contracting surface slowly creep round the shoot, deserting by slow degrees the southern side and encroaching on the eastern side, and so round by the north, by the west, again to the south; in this case the shoot would remain always bowed with the painted line appearing on the convex, on the lateral, and concave surfaces, and with the point of the shoot successively directed to all points of the compass. In fact, we should then have the exact kind of movement seen in the revolving shoots of twining plants. I have spoken in the illustration, for brevity's sake, of the cells along each face successively contracting; of course turgescence of the cells on the opposite face, or both forces combined, would do equally well.

It must not be supposed that the revolving movement of twining plants is as regular as that given in this illustration; in very many cases the tip describes an ellipse, even a very narrow ellipse. To recur once again to our illustration, if we suppose the southern and then the northern face of the sapling to contract, the summit would describe a simple arc; if the contraction first travelled a very little to the eastern face, and during the return a very little to the western face, a narrow ellipse would be described; and the sapling would become straight as it passed to and fro by the central point. A complete straightening of the shoot may often be observed in revolving plants; but the weight of the shoot apparently interferes with the regularity of the movement, and with the place of straitening. The movement is often (in appearance at least) as if the southern, eastern, and northern faces had contracted, but not the western face; so that a semicircle is described, and the shoot becomes straight and upright in one part of its course.

When a revolving shoot consists of several internodes, the several lower ones bend together at the same rate, but the one or two terminal internodes bend at a slower rate; hence, though at times all the internodes may be bowed in the same line, at other times the shoot is rendered slightly serpentine, as I have often observed. The rate of revolution of the whole shoot, if judged by the movement of the extreme tip, is thus at times accelerated and retarded. One other point must be noticed. Authors have observed that the end of the shoot in many twining plants is completely hooked; this is very general, for instance, with the Asclepiadaceæ. The hooked tip, in all the cases which I observed, viz. in *Ceropegia, Sphærostema, Clerodendron, Wistaria, Stephania, Akebia,* and *Siphomeris,* has exactly the same kind of movement as the other revolving internodes; for a line painted on the convex surface becomes lateral and then concave; but, owing to the youth of these terminal internodes, the reversal of the hook is a slower process than the revolving movement. This strongly marked tendency in the young terminal and flexible internodes to bend more abruptly than the other internodes is of service to the plant; for not only does the hook thus formed sometimes serve to catch a support, but (and this seems to be much more important) it causes the extremity of the shoot to embrace much more closely its support than it otherwise could have done, and thus aids in preventing the stem from being blown away from it during windy weather, as I have many times observed. In *Lonicera brachypoda* the hook only straightened itself periodically, and never became reversed. I will not assert that the tips of all twining plants, when hooked, move as above described; for this position may in some cases be due to the manner of growth, as with the bent tips of the shoots of the common vine, and more plainly with those of *Cissus discolor*; these plants, however, are not spiral twiners.

The purpose of the spontaneous revolving movement, or, more strictly speaking, of the continuous bending movement successively directed to all points of the compass, is, as Mohl has remarked, obviously in part to favour the shoot finding a support. This is admirably effected by the revolutions carried on night and day, with a wider and wider circle swept as the shoot increases in length. But as we now understand the nature of the movement, we can see that, when at last the shoot meets with a support, the motion at the point of contact is necessarily arrested, but the free projecting part goes on revolving. Almost immediately another and upper point of the shoot is brought into contact with the support and is arrested; and so onwards to the extremity of the shoot; and thus it winds round its support. When the shoot follows the sun in its revolving course, it winds itself round the support from right to left, the support being supposed to stand in front of the beholder; when the shoot revolves in an opposite direction, the line of winding is reversed. As each internode loses from age its power of revolving, it loses its power of spirally twining round a support. If a man swings a rope round his head, and the end hits a stick, it will coil round the stick according to the direction of the swinging rope; so it is with twining plants, the continued contraction or turgescence of the cells along the free part of the shoot replacing the momentum of each atom of the free end of the rope.

All the authors, except Von Mohl, who have discussed the

spiral twining of plants maintain that such plants have a natural tendency to grow spirally. Mohl believes (S. 112) that twining stems have a dull kind of irritability, so that they bend towards any object which they touch. Even before reading Mohl's interesting treatise, this view seemed to me so probable that I tested it in every way that I could, but always with negative results. I rubbed many shoots much harder than is necessary to excite movement in any tendril or in any foot-stalk of a leaf-climber, but without result. I then tied a very light forked twig to a shoot of a Hop, a *Ceropegia*, *Sphærostema*, and *Adhatoda*, so that the fork pressed on one side alone of the shoot and revolved with it; I purposely selected some very slow revolvers, as it seemed most likely that these would profit from possessing irritability; but in no case was any effect produced. Moreover, when a shoot winds round a support, the movement is always slower, as we shall immediately see, than whilst its revolves freely and touches nothing. Hence I conclude that twining stems are not irritable; and indeed it is not probable that they should be so, as nature always economizes her means, and irritability would be superfluous. Nevertheless I do not wish to assert that they are never irritable; for the growing axis of the leaf-climbing, but not spirally twining, *Lophospermum scandens* is, as we shall hereafter see, certainly irritable; but this case gives me confidence that ordinary twiners do not possess this quality, for directly after putting a stick to the *Lophospermum*, I saw that it behaved differently from any true twiner or any other leaf-climber.

The belief that twiners have a natural tendency to grow spirally probably arose from their assuming this form when wound round a support, and from the extremity, even whilst remaining free, sometimes assuming this same form. The free internodes of vigorously growing plants, when they cease to revolve, become straight, and show no tendency to be spiral; but when any shoot has nearly ceased to grow, or when the plant is unhealthy, the extremity does occasionally become spiral. I have seen this in a remarkable degree with the ends of the shoots of the *Stauntonia* and of the allied *Akebia*, which became closely wound up spirally, just like a tendril, especially after the small, ill-formed leaves had perished. The explanation of this fact is, I believe, that the lower parts of such terminal internodes very gradually and successively lose their power of movement, whilst the portions just above move onwards, and in their turn become motionless; and this ends in forming an irregular spire.

When a revolving shoot strikes a stick, it winds round it rather more slowly than it revolves. For instance, a shoot of the *Ceropegia* took 9 h. 30 m. to make one complete spire round a stick, whilst it revolved in 6 h.; *Aristolochia gigas* revolved in about 5 h., but took 9 h. 15 m. to complete its spire. This, I presume, is due to the continued disturbance of the moving force by its arrestment at each successive point; we shall hereafter see that even shaking a plant retards the revolving movement. The terminal internodes of a long, much-inclined, revolving shoot of the *Ceropegia*, after they had wound round a stick, always slipped up it, so as to render the spire more open than it was at first; and this was evidently due to the force which caused the revolutions being now almost freed from the constraint of gravity, and allowed to act freely. With the *Wistaria*, on the other hand, a long horizontal shoot wound itself at first in a very close spire, which remained unchanged; but subsequently, as the shoot grew, it made a much more open spire. With all the many plants which were allowed freely to ascend a support, the terminal internodes made at first a close spire; and this, during windy weather, well served to keep the shoots in contact with their support; but as the penultimate internodes grew in length, they pushed themselves up for a considerable space (ascertained by coloured marks on the shoot and on the support) round the stick, and the spire became more open.

It follows from this latter fact that the position occupied by each leaf with respect to the support, in fact, depends on the growth of the internodes after they have become spirally wound round it. I mention this on account of an observation by Palm (S. 34), who states that the opposite leaves of the Hop always stand exactly over each other, in a row, on the same side of the supporting stick, though this may differ in thickness. My sons visited a hop-field for me, and reported that though they generally found the points of insertion of the leaves over each other for a space of two or three feet in height, yet this never occurred up the whole length of a pole, the point of insertion forming, as might have been expected, an irregular spire. Any irregularity in the pole entirely destroyed the regularity of position of the leaves. From casual inspection, it appeared to me that the opposite leaves of *Thunbergia alata* were arranged in a line up the sticks round which they had twined; accordingly I raised a dozen plants, and gave them sticks of various thicknesses and string to twine round; and in this case one alone out of the dozen had its leaves arranged in a perpendicular line: so I conclude that there is nothing remarkable in Palm's statement.

The leaves of twining-plants rise from the stem (before it has twined) either alternately, or oppositely, or in a spire; in this latter case the line of insertion of the leaves and the course of revolution or of twining coincide. This fact has been well shown by Dutrochet *, who found different individuals of *Solanum Dulcamara* twining in opposite directions, and these had their leaves spirally arranged in opposite directions. A dense whorl of many leaves would apparently be incommodious for a twining plant, and some authors have supposed that none have their leaves thus arranged; but a twining *Siphomeris* has whorls of three.

If a stick which has arrested a revolving shoot, but has not as yet been wound round, be suddenly taken away, the shoot generally springs forward, showing that it has continued to press against the stick. If the stick, shortly after having been wound round, be withdrawn, the shoot retains for a time its spiral form, then straightens itself, and again commences to revolve. The long, much-inclined shoot of the *Ceropegia* previously alluded to offered some curious peculiarities. The lower and older internodes, which continued to revolve, had become so stiff that they were incapable, on repeated trials, of twining round a thin stick, showing that the power of movement was retained after flexibility had been lost. I then moved the stick to a greater distance, so that it was struck by a point $2\frac{1}{2}$ inches from the extremity of the penultimate internode; and it was then neatly wound round by this part and by the ultimate internode. After leaving the spirally wound shoot for eleven hours, I quietly withdrew the stick, and in the course of the day the curled part straightened itself and recommenced revolving; but the lower and not curled portion of the penultimate internode did not move, a sort of hinge separating the moving and the motionless part of the same internode. After a few days, however, I found that the lower part of this internode had likewise recovered its revolving power. These several facts show that, in the arrested portion of a revolving shoot, the power of movement is not immediately lost, and that when temporarily lost it can be recovered. When a shoot has remained for a considerable time wound round its support, it permanently retains its spiral form even when the support is removed.

* Comptes Rendus, 1844, tom. xix. p. 295, and Annales des Soc. Nat. 3rd series, Bot., tom. ii. p. 163.

When a stick was placed so as to arrest the lower and rigid internodes of the *Ceropegia* at the distance at first of 15 and then of 21 inches from the centre of revolution, the shoot slowly and gradually slid up the stick, so as to become more and more highly inclined; and then, after an interval sufficient to have allowed of a semirevolution, it suddenly bounded from the stick and fell over to the opposite side, to its ordinary slight inclination. It now recommenced revolving in its usual course, so that after a semirevolution it again came into contact with the stick, again slid up it, and again bounded from it. This movement of the shoot had a very odd appearance, as if it were disgusted with its failure but resolved to try again. We shall, I think, understand this movement by considering the former illustration of the sapling, in which the contracting surface was supposed to creep from the southern, by the eastern, to the northern, and thence back again by the western side to the southern face, successively bowing the sapling in all directions. Now with the *Ceropegia*, the stick being placed a very little to the east of due south of the plant, the eastern contraction could produce no effect beyond pressing the rigid internode against the stick; but as soon as the contraction on the northern face began, it would slowly drag the shoot up the stick; and then, as soon as the western contraction had well begun, the shoot would be drawn from the stick, and its weight, coinciding with the north-western contraction, would cause it suddenly to fall to the opposite side with its proper slightly inclined positions; and the ordinary revolving movement would go on. I have described this case because it first made me understand the order in which the contracting or turgescent cells of revolving shoots must act.

The view just given further explains, as I believe, a fact observed by Von Mohl (S. 135), namely, that a revolving shoot, though it will twine round an object as thin as a thread, cannot do so round a thick support. I placed some long revolving shoots of a *Wistaria* close to a post between 5 and 6 inches in diameter, but they could not, though aided by me in many ways, wind round it. This apparently is owing to the flexure of the shoot, when winding round an object so gently curved as this post, not being sufficient to hold the shoot to its place when the contracting force creeps round to the opposite surface of the shoot; so that it is at each revolution withdrawn from its support.

When a shoot has grown far beyond its support, it sinks downwards from its weight, as already explained in the case of the

Hop, with the revolving end always turning upwards. If the support be not lofty, it falls to the ground, and, resting there, the extremity rises again. Sometimes several shoots, when flexible, twine together into a cable, and thus support each other. Single thin depending shoots, such as those of the *Sollya Drummondii*, will turn abruptly back and wind upwards on themselves. The greater number of the depending shoots, however, of one twining plant, the *Hibbertia dentata*, showed but little tendency to turn upwards. In other cases, as with the *Cryptostegia grandiflora*, several internodes which at first were flexible and revolved, if they did not succeed in twining round a support, became quite rigid, and, supporting themselves upright, carried on their summit the younger revolving internodes.

Here will be a convenient place to give a Table showing the direction and rate of movement of several twining plants, with a few appended remarks. These plants are arranged according to Lindley's 'Vegetable Kingdom' of 1853; and they have been selected from all parts of the series to show that all kinds behave in a nearly uniform manner*.

Twining plants not aided by tendrils or by irritable leaf-stalks.

(ACOTYLEDONS.)

Lygodium scandens (Polypodiaceæ) moves against the sun.

	h. m.				h. m.	
June 18, 1st circle	6	0	[ing).	June 19, 4th circle	5	0 (very hot day).
„ 18, 2nd „	6	15 (late in even-		„ 20, 5th „	6	0
„ 19, 3rd „	5	32 (very hot day).				

Lygodium articulatum moves against the sun.

	h. m.				h. m.	
July 19, 1st circle	16	30 (shoot very	July 21, 3rd circle	......	8	0
„ 20, 2nd „	15	0 [young).	„ 22, 4th „	......	10	30

(MONOCOTYLEDONS.)

Ruscus androgynus (Liliaceæ), placed in the hot-house, moves against the sun.

	h. m.				h. m.	
May 24, 1st circle	6	14 (shoot very	May 26, 5th circle	......	2	50
„ 25, 2nd „	2	21 [young).	„ 27, 6th „	......	3	52
„ 25, 3rd „	3	37	„ 27, 7th „	......	4	11
„ 25, 4th „	3	22				

---

* I am much indebted to Dr. Hooker for having sent me many plants from Kew; and to Mr. Veitch, of the Royal Exotic Nursery, for having generously given me a large collection of fine specimens of climbing plants. Professor Asa Gray, Prof. Oliver, and Dr. Hooker have afforded me, as on many previous occasions, much information and many valuable references.

(MONOCOTYLEDONS, *continued.*)

Asparagus (unnamed species from Kew) (Liliaceæ) moves against the sun, placed in hothouse.

	h. m.
Dec. 26, 1st circle	5 0
„ 27, 2nd „	5 40

Tamus communis (Dioscoreaceæ). A young shoot from a potted tuber placed in the greenhouse; follows the sun.

	h. m.			h. m.
July 7, 1st circle	3 10	July 8, 4th circle	2 56
„ 7, 2nd „	2 38	„ 8, 5th „	2 30
„ 8, 3rd „	3 5	„ 8, 6th „	2 30

Lapagerea rosea (Philesiaceæ), in greenhouse, follows the sun.

	h. m.
March 9, 1st circle	26 15 (shoot young).
„ 10, semicircle	8 15
„ 11, 2nd circle	11 0
„ 12, 3rd „	15 30
„ 13, 4th „	14 15
„ 16, 5th „	8 40 when placed in the hothouse; but the

next day the shoot remained stationary.

Roxburghia viridiflora (Roxburghiaceæ) moves against the sun; it travelled a circle in about 24 hours.

(DICOTYLEDONS.)

Humulus Lupulus (Urticaceæ) follows the sun.

	h. m.			h. m.
April 9, 2 circles	4 16	August 14, 6th circle	2 2
Aug. 13, 3rd circle	2 0	„ 14, 7th „	2 0
„ 14, 4th „	2 20	„ 14, 8th „	2 4
„ 14, 5th „	2 16		

A plant placed in a room; a semicircle was performed in travelling from the light in 1 h. 33 m., in travelling to the light in 1 h. 13 m.: difference of rate 20 m.

Akebia quinata (Lardizabalaceæ), placed in hothouse, moves against the sun.

	h. m.			h. m.
March 17, 1st circle 4 0 (shoot		March 18, 3rd circle	1 30
„ 18, 2nd „ 1 40 [young).		„ 19, 4th „	1 45

Stauntonia latifolia (Lardizabalaceæ), placed in hothouse, moves against the sun.

	h. m.
March 28, 1st circle	3 30
„ 29, 2nd „	3 45

Sphærostema marmoratum (Schizandraceæ) follows the sun.

August 5th, 1st circle in about 24 h.; 2nd circle in 18 h. 30 m.

Stephania rotunda (Menispermaceæ) moves against the sun.

	h. m.			h. m.
May 27, 1st circle	5 5	June 2, 3rd circle	5 15
„ 30, 2nd „	7 6	„ 3, 4th „	6 28

(DICOTYLEDONS, *continued.*)

Thryallis brachystachya (Malpighiaceæ) moves against the sun: one shoot made a circle in 12 h., and another in 10 h. 30 m.; but the next day, which was much colder, the first shoot in my study took 10 h. to perform only a semicircle.

Hibbertia dentata (Dilleniaceæ), placed in the hothouse, followed the sun, and made (May 18th) a circle in 7 h. 20 m.; on the 19th, reversed its course and moved against the sun, and made a circle in 7 h.; on the 20th, moved against the sun one-third of circle, and then stood still; on the 26th, followed the sun for two-thirds of circle, and then returned to its starting-point, taking for this double course 11 h. 46 m.

Sollya Drummondii (Pittosporaceæ) moves against the sun; in greenhouse.

	h. m.		h. m.
April 4, 1st circle	4 25 [day.)	April 6, 3rd circle	6 25
„ 5, 2nd „	8 0 (very cold	„ 7, 4th „	7 5

Polygonum dumetorum (Polygonaceæ). This case is taken from Dutrochet (p. 299), as I observed no allied plant; follows the sun. Three shoots cut off and placed in water made circles in 3 h. 10 m., 5 h. 20 m., and 7 h. 15 m.

Wistaria Chinensis (Leguminosæ), in greenhouse, moves against the sun.

	h. m.		h. m.
May 13, 1st circle	3 5	May 24, 4th circle	3 21
„ 13, 2nd „	3 20	„ 25, 5th „	2 37
„ 16, 3rd „	2 5	„ 25, 6th „	2 35

Phaseolus vulgaris (Leguminosæ), in greenhouse, moves against the sun.

	h. m.
May, 1st circle	2 0
„ 2nd „	1 55
„ 3rd „	1 55

Dipladenia urophylla (Apocynaceæ) moves against the sun.

	h. m.
April 18, 1st circle	8 0
„ 19, 2nd „	9 15
„ 30, 3rd „	9 40

Dipladenia crassinoda moves against the sun.

	h. m.
May 16, 1st circle	9 5
July 20, 2nd „	8 0
„ 21, 3rd „	8 5

Ceropegia Gardnerii (Asclepiadaceæ) moves against the sun.

		h. m.
Shoot very young, 2 inches in length.	1st circle in	7 55
Shoot still young	2nd „	7 0
Long shoot	3rd „	6 33
Long shoot	4th „	5 15
Long shoot	5th „	6 45

Stephanotis floribunda (Asclepiadaceæ) moves against the sun, and made a circle in 6 h. 40 m., a second circle in about 9 hours.

(DICOTYLEDONS, *continued.*)

Hoya carnosa (Asclepiadaceæ) made several circles in from 16 h. to 22 h. or 24 h.

Convolvulus major (Convolvulaceæ) moves against the sun. Plant placed in room with lateral light.

1st circle ... 2 h. 42 m. $\left\{ \begin{array}{l} \text{Semicircle, from light in 1 h. 14 m., to light} \\ \text{1 h. 28 m. : difference 14 m.} \end{array} \right.$

2nd circle... 2 h. 47 m. $\left\{ \begin{array}{l} \text{Semicircle, from light in 1 h. 17 m., to light} \\ \text{1 h. 30 m. : difference 13 m.} \end{array} \right.$

Convolvulus sepium (large-flowered cultivated var.) moves against the sun. Two circles, each in 1 h. 42 m. : difference in semicircle from and to light 14 m.

Ipomæa jucunda (Convolvulaceæ) moves against the sun, placed in my study, with windows facing the north-east. Weather hot.

1st circle 5 h. 30 m. .... $\left\{ \begin{array}{l} \text{Semicircle, from light in 4 h. 30 m., to light} \\ \text{1 h. 0 m. : difference 3 h. 30 m.} \end{array} \right.$

2nd circle 5 h. 20 m. (Late in afternoon : circle completed at 6 m. 40 h. P.M.) $\left. \begin{array}{l} \text{Semicircle, from light in 3 h. 50 m., to light} \\ \text{1 h. 30 m. : difference 2 h. 20 m.} \end{array} \right.$

We have here a remarkable instance of the power of light in retarding and hastening the revolving movement.

Rivea tiliæfolia (Convolvulaceæ) moves against the sun, and made four revolutions in 9 h. ; so that each, on average, was performed in 2 h. 15 m.

Plumbago rosea (Plumbaginaceæ) follows the sun. The shoot did not begin to revolve until nearly a yard in height ; it then made a fine circle in 10 h. 45 m. During the next few days it continued to move, but irregularly. On August 15th the shoot followed, during a period of 10 h. 40 m., a long and deeply zigzag course and then made a broad ellipse. The figure thus traced altogether apparently represented three ellipses, each of which averaged 3 h. 33 m. for its completion.

Jasminum pauciflorum, Bentham (Jasminaceæ), moves against the sun. First circle in 7 h. 15 m., second circle rather more quickly.

Clerodendrum Thomsonii (Verbenaceæ) follows the sun.

	h. m.	
April 12, 1st circle	5 45	(shoot very young).
„ 14, 2nd „	3 30	
„ 18, semicircle	5 0	(directly after the plant was shaken in
„ 19, 3rd circle	3 0	[being moved).
„ 20, 4th „	4 20	

Tecoma jasminoides (Bignoniaceæ) moves against sun.

h. m.	h. m.
March 17, 1st circle 6 30	March 22, 3rd circle 8 30 (very cold
„ 19, 2nd „ 7 0	„ 24, 4th „ 6 45 [day).

Thunbergia alata (Acanthaceæ) moves against sun.

h. m.	h. m.
April 14, 1st circle 3 20	April 18, 3rd circle 2 55 [noon).
„ 18, 2nd „ 2 50	„ 18, 4th „ 3 55 (late in after-

Adhadota cydonæfolia (Acanthaceæ) follows the sun. A young shoot made

C

(DICOTYLEDONS, *continued*.)

a semicircle in 24 h.; subsequently made a circle in between 40 h. and 48 h.; subsequently did not complete a circle in 50 h. Another shoot, however, made a circle in 26 h. 30 m.

Mikania scandens (Compositæ) moves against the sun.

		h. m.
March 14, 1st circle		3 10
„ 15, 2nd	„	3 0
„ 16, 3rd	„	3 0
„ 17, 4th	„	3 33
April 7, 5th	„	2 50
„ 7, 6th	„	2 40 { This circle was made after a copious intentional watering with cold water at 47° Fahr.

Combretum argenteum (Combretaceæ) moves against the sun.

	h. m.	
Jan. 24, 1st circle..............	2 55 {	Early in morning, when the temperature of the house had fallen a little.
„ 24, 2 circles, each at an average of	2 20 / 2 20	
„ 25, 4th circle..............	2 25	

Combretum purpureum revolves not quite so quickly as *C. argenteum*.

Loasa aurantiaca (Loasaceæ). First plant moved against the sun.

	h. m.		h. m.
June 20, 1st circle......	2 37	June 21, 4th circle......	2 35
„ 20, 2nd „	2 13	„ 22, 5th „	3 26
„ 20, 3rd „	4 0	„ 23, 6th „	3 5

Second plant followed the sun.

	h. m.	
July 11, 1st circle...........	1 51	
„ 11, 2nd „	1 46	
„ 11, 3rd „	1 41 } Very hot day.
„ 11, 4th „	1 48	
„ 12, 5th „	2 35	Cool morning.

Scyphanthus elegans (Loasaceæ) follows the sun.

	h. m.		h. m.
June 13, 1st circle......	1 45	June 14, 4th circle......	1 59
„ 13, 2nd „	1 17	„ 14, 5th „	2 3
„ 14, 3rd „	1 36		

Siphomeris or *Lecontea* (unnamed sp.) (Cinchonaceæ) follows the sun.

	h. m.	
May 25, semicircle	10 27	(shoot extremely young).
„ 26, 1st circle	10 15	(shoot still young).
„ 30, 2nd „	8 55	
June 2, 3rd „	8 11	
„ 6, 4th „	6 8	
„ 8, 5th „	7 20 } Taken from the hothouse and placed
„ 9, 6th „	8 36 } in a room in my house.

Manettia bicolor (Cinchonaceæ), young plant, follows the sun.

	h. m.
July 7, 1st circle.....................	6 18
„ 8, 2nd „	6 53
„ 9, 3rd „	6 30

(DICOTYLEDONS, *continued.*)

Lonicera brachypoda (Caprifoliaceæ) follows the sun, in a warm room in
the house.

<pre>
                         h. m.
April, 1st circle about 9 10
  „   2nd „   „   12 20 (another shoot very young).
  „   3rd „   „   7 30
                            ⎧ In this latter circle, the semicircle from the
  „   4th „   „   8 0⎨   light took 5 h. 23 m., and to the light
                            ⎩   2 h. 37 m. : difference 2 h. 46 m.
</pre>

Aristolochia gigas (Aristolochiaceæ) moves against the sun.

<pre>
                            h. m.
July 22, 1st circle......... 8 0 (rather young shoot).
  „ 23, 2nd „ 7 15
  „ 24, 3rd „   .... 5 0 (about).
</pre>

In the foregoing table, which includes twining plants belonging
to as widely different orders as is possible, we see that the contraction or turgescence of the cells circulating round the axis, on
which the revolving movement depends, differs much in rate. As
long as a plant remains under the same conditions, the rate is
often remarkably uniform, as we see with the Hop, *Mikania*,
Phaseolus, &c. The *Scyphanthus* made one revolution in 1 h. 17 m.,
and this is the quickest rate observed; but we shall afterwards
see a tendril-bearing Passiflora revolving even more rapidly. A
shoot of the *Akebia quinata* made a revolution in 1 h. 30 m., and
three revolutions at the average rate of 1 h. 38 m.; a Convolvulus
made two revolutions at the average of 1 h. 42 m., and *Phaseolus
vulgaris* three at the average of 1 h. 57 m. On the other hand, some
plants take 24 h. for a single revolution, and the *Adhadota* sometimes required 48 h.; yet this latter plant is an efficient twiner.
Species of the same genus move at different rates. The rate does
not seem governed by the thickness of the shoots: those of the
Sollya are as thin and flexible as string, but move slower than the
thick and fleshy shoots of the *Ruscus*, which seems so little fitted
for movement of any kind; the shoots of the *Wistaria*, which become woody, move faster than those of the *Ipomœa* or *Thunbergia*.

We know that the internodes, whilst very young, do not acquire their proper rate of movement; hence several shoots on the
same plant may sometimes be seen revolving at different rates.
The two or three, or even more, internodes which are first formed
above the cotyledons, or above the perennial root-stock, do not
move; these first-formed shoots can support themselves, and
nothing superfluous is granted them.

A greater number of twiners revolve in a course opposed to
that of the sun, or to the hands of a watch, than in the reversed

course, and, consequently, the majority, as is well known, ascend
their supports from left to right. Occasionally, though rarely,
plants of the same order twine in opposite directions, of which
Mohl (S. 125) gives a case in the Leguminosæ and we have in
the table another in the Acanthaceæ. At present no instance is
known of two species of the same genus twining in opposite directions; and this is a singular fact, because different individuals
of *Solanum dulcamara* (Dutrochet, tom. xix. p. 299) revolve and
twine in both directions: this plant, however, is a most feeble
twiner. *Loasa aurantiaca* (Léon, p. 351) offers a much more
striking case: I raised seventeen plants: of these eight revolved
in opposition to the sun, and ascended from left to right; five
followed the sun, and ascended from right to left; and four revolved and twined first in one direction, and then reversed their
course*, the petioles of the opposite leaves affording a *point
d'appui* for the reversal of the spire. One of these four plants
made seven spiral turns from right to left, and five turns from
left to right. These individuals of the *Loasa* are interesting, as
showing how almost every change is effected most gradually.
For another plant in the same family, the *Scyphanthus elegans*,
habitually twines in this manner. I raised many plants of it, and
the stems of all took one turn, or occasionally two or even three
turns in one direction, and then, ascending for a short space straight,
reversed their course and took one or two turns in an opposite
direction. The reversal of the curvature occurred at any point in
the stem, even in the middle of an internode. Had I not seen
this case, I should have thought its occurrence most improbable.
It could hardly occur with any plant which ascended above a few
feet in height, or which lived in an exposed situation; for the
stem could be easily pulled from its support with little unwinding;
nor could it have adhered at all, had not the internodes soon become moderately rigid. With leaf-climbers, as we shall soon see,
analogous cases frequently occur; but these present no difficulty,
as the stem is secured by the clasping petioles.

In the many other revolving and twining plants observed by
me, I never but twice saw the movement reversed; once, and only
for a short space, in *Ipomœa jucunda*; but frequently with *Hibbertia dentata*. This plant at first much perplexed me, for I continually observed its long and flexible shoots, evidently well fitted
for twining, make a whole or half or quarter circle in one direction

* I raised nine plants of the hybrid *Loasa Herbertii*, and six of these reversed their spire in ascending their supports.

and then in the opposite direction; consequently, when I placed the shoots near thin or thick sticks, or stretched string, they seemed perpetually to be trying to ascend these supports, but failed. I then surrounded the plant with a mass of branched twigs; the shoots ascended, and passed through them, but several came out laterally, and their depending extremities seldom turned upwards as is usual with twining plants. Finally, I surrounded another plant with many thin upright sticks, and placed this plant near the other plant with the twigs; and now the *Hibbertia* had got what it liked, for it twined up the parallel sticks, sometimes winding round one and sometimes round several; and the shoots travelled laterally from one to the other plant; but as the plants grew older, some of the shoots twined regularly up a thin upright stick. Though the revolving movement was sometimes in one direction and sometimes in the other, the twining was invariably from left to right; so that the more potent or persistent movement of revolution must have been in opposition to the course of the sun. It would appear that this *Hibbertia* is adapted to ascend by twining, and to ramble laterally over the thick Australian scrub.

I have described this case in some detail, because, as far as I have seen, it is rare to find with twining plants any especial adaptations, in which respect they differ much from the more highly organized tendril-bearers. The *Solanum dulcamara*, as we shall presently see, can twine only round such stems as are both thin and flexible. Most twining plants apparently are adapted to ascend supports of different thicknesses. Our English twiners, as far as I have seen, never twine round trees, excepting the Honeysuckle (*Lonicera periclymenum*), which I have observed twining up a young beech-tree nearly 4½ inches in diameter. Mohl (S. 134) found that the *Phaseolus multiflorus* and *Ipomæa purpurea* could not, when placed in a room with the light entering on one side, twine round sticks between 3 and 4 inches in diameter; for this interfered, in a manner presently to be explained, with the revolving movement. In the open air, however, the *Phaseolus* twined round a support of the above thickness, but failed in twining round one 9 inches in diameter. Nevertheless, some twiners of the warmer temperate regions can manage this latter degree of thickness; for I hear from Dr. Hooker that at Kew the *Ruscus androgynus* ascends a column 9 inches in diameter; and although a *Wistaria* grown by me in a small pot tried in vain for weeks to get round a post between 5 and 6 inches in thickness, yet at Kew a plant ascended a trunk above 6 inches in diameter. The tropical twiners, on the other hand, can ascend thick trees. I hear from Drs. Thomson and Hooker that this is the case with the *Butea parviflora*, one of the Menispermaceæ, and with some Dalbergias and other Leguminosæ. This power would evidently be almost necessary for twining plants inhabiting tropical forests, as otherwise they could hardly ever reach the light. In our temperate countries twining plants which die down every year to the root would suffer if they were enabled to twine round trunks of trees, for they could not grow tall enough in a single season to reach the summit and gain the light.

By what means some twining plants are adapted to ascend only thin stems, whilst others can twine round thick trees, I do not know. It appeared to me probable that twining plants with very long revolving shoots might be able to ascend thick supports; accordingly I placed *Ceropegia Gardnerii* near a post 6 inches in diameter, but the shoots entirely failed to wind round it; their length and power of movement apparently serving merely to find some distant but thin stem round which to twine. The *Sphæro-stemma marmoratum* is a vigorous tropical twiner, and as it is a very slow revolver, I thought that this latter circumstance might aid it in ascending a thick support; but though it was able to wind round the 6-inch post, it could do this only on the same level or plane, and could not ascend in a spire. We can, however, see, in accordance with the views previously explained, that a revolving shoot, which, after coming into contact with any support, quickly lost its power of movement, would not again be drawn away from its support by the returning or opposite movement, and therefore remaining in contact with it, might thus ascend a thick support. But whether this slight difference in retaining for some time or in quickly losing the power of movement after coming into contact with a support alone determines how thick an object the stem can ascend I do not know.

As ferns differ so much from phanerogamic plants, it may be worth while here to show that twining ferns act in no respect differently from other twining plants. In *Lygodium articulatum* the two internodes first formed above the root-stock did not move; the third from the ground revolved, and at first very slowly. This species is a slow revolver: but *L. scandens* made five revolutions at an average rate of 5 h. 45 m.; and this represents fairly well the usual rate, taking quick and slow movers, amongst phanerogamic plants. The rate was accelerated by increased temperature. The two young upper internodes alone moved. A line painted along the surface of a revolving internode which was at the time convex, became first lateral, then concave, and ultimately convex again. Neither the internodes nor petioles are irritable when rubbed. The movement is in the more usual direction, namely in opposition to the course of the sun; and when the stem has twined round a thin stick, it becomes twisted on its own axis in the same direction. After the young internodes have twined round a stick, their continued growth causes them to slip a little upwards and onwards. If the stick be soon removed, the internodes straighten themselves, and recommence revolving. The extremities of the depending shoots turn upwards, and twine on themselves. In all these respects we have complete identity with phanerogamic twining plants; and the above enumeration may serve as a summary of the leading characteristics of common twining plants.

The power of revolving depends on the general health and vigour of the plant, as has laboriously been shown to be the case by Palm. But the movement of each separate internode is so independent of the others, that cutting off an upper one does not affect the revolutions of a lower one. When, however, Dutrochet cut off two whole shoots of the Hop, and placed them in water, the movement was greatly retarded; for one revolved in 20 h. and the other in 23 h., whereas they ought to have revolved in between 2 h. and 2 h. 30 m. Cut shoots of the Kidney-bean were similarly retarded, but in a less degree. I have repeatedly observed that carrying a plant from the greenhouse to my house, or from one to another part of the greenhouse, always stopped the movement for a time; hence I conclude that naturally exposed plants would not make their revolutions during stormy weather. A decrease in temperature always caused a considerable retardation in the rate of revolution; but Dutrochet (tom. xvii. pp. 994, 996) has given such precise observations on this head with respect to the tendril-bearing Pea that I need say nothing more. When twining plants are placed near a window in a room, the light in some cases has a remarkable power (as was likewise observed by Dutrochet, p. 998, with the Pea) on the revolving movement, but different in degree with different plants: thus *Ipomœa jucunda* (as may be seen in the table) revolved in 5 h. 20 m., the semicircle from the light taking 4 h. 30 m., and that towards the light only 1 h.; *Lonicera brachypoda* revolved, in a reversed direction to the *Ipomœa*, in 8 h., the semicircle from the light taking 5 h. 23 m., and that to the light only 2 h. 37 m. From the rate of revolution in all the plants which I have observed being nearly the same during the night and the day, I infer that the action of the light is confined to retarding one semicircle and accelerating the other, so as not to greatly modify the whole rate. This action is remarkable when we reflect how little the leaves are developed on the young and very thin revolving internodes. It is the more remarkable, as botanists have thought (Mohl, S. 119) that twining plants are but little sensitive to the action of light.

I will conclude my account of twining plants by collecting a few miscellaneous and curious cases. With most twining plants all the branches, however many there may be, go on revolving together; but, according to Mohl (S. 4), the main stem of *Tamus elephantipes* does not twine—only the branches. On the other hand, with the Asparagus, given in the table, the leading shoot alone, and not the branches, revolved and twined; but it should be stated that the plant was not growing vigorously. My plants of *Combretum argenteum* and *C. purpureum* made nume-rous short healthy shoots; but they showed no signs of revolv-ing, and I could not conceive how these plants could be climbers; but at last *C. argenteum* put forth from the lower part of one of its main branches a thin shoot, 5 or 6 feet in length, differing greatly in appearance from the previous shoots from its leaves being little developed, and this shoot revolved vigorously and twined. So that this plant produces shoots of two sorts. With *Periploca Græca* (Palm, S. 43) the uppermost shoots alone twine. *Polygonum convolvulus* twines only during the middle of the sum-mer (Palm. S. 43, 94): plants growing vigorously in the autumn show no inclination to twine. The majority of Asclepiadaceæ are twiners; but *Asclepias nigra* only "in fertiliori solo incipit scan-dere sub volubili caule" (Willdenow, quoted and confirmed by Palm, S. 41). *Asclepias vincetoxicum* does not regularly twine, but only occasionally (Palm, S. 42; Mohl, S. 112) when growing under certain conditions. So it is with two species of *Ceropegia*, as I hear from Prof. Harvey, for these plants in their native dry South African home generally grow erect, from 6 inches to 2 feet in height, a very few taller specimens showing some inclination to curve; but when cultivated near Dublin, they regularly twined up sticks 5 or 6 feet in height. Most Convolvulaceæ are excellent twiners; but *Ipomœa argyræoides* in South Africa almost always grows erect and compact, from about 12 to 18 inches in height, one specimen alone in Prof. Harvey's collection showing an evident disposition to twine. Seedlings, on the other hand, raised near

Dublin twined up sticks above 8 feet in height. These facts are highly remarkable; for there can hardly be a doubt that in the dryer provinces of South Africa these plants must have propagated themselves for thousands of generations in an erect condition; and yet during this whole period they have retained the innate power of spontaneously revolving and twining, whenever their shoots become elongated under proper conditions of life. Most of the species of *Phaseolus* are twiners; but certain varieties of the *P. multiflorus* produce (Léon, p. 681) two kinds of shoots, some upright and thick, and others thin and twining. I have seen striking instances of this curious case of variability with " Fulmer's dwarf forcing-bean," on which occasionally a long twining shoot appeared.

Solanum dulcamara is one of the feeblest and poorest of twiners : it may often be seen growing as an upright bush, and when growing in the midst of a thicket merely scrambles up the branches without twining; but when, according to Dutrochet (tom. xix. p. 299), it grows near a thin and flexible support, such as the stem of a nettle, it twines round it. I placed sticks round several plants and vertical stretched strings close to others, and the strings alone were ascended by twining. We here, perhaps, see the first stage in the habit of twining; and the stem twines indifferently to the right or the left. Some other species of the genus, and of another genus, viz. *Habrothamnus*, of the same family of Solanaceæ, which are described in horticultural works as twining plants, seemed to possess this faculty in a very feeble manner. On the other hand, I suspect that with *Tecoma radicans* we have the last vestige of a lost habit: this plant belongs to a group abounding with twining and with tendril-bearing species, but it ascends by rootlets like those of the Ivy; yet I observed that the young internodes seldom remained quite stationary, but performed slight irregular movements which could hardly be accounted for by changes in the action of the light. Anyhow it need not be supposed that there would be any difficulty in the passage from a spirally twining plant to a simple root-climber; for the young internodes of *Bignonia Tweedyana* and of *Hoya carnosa* revolve and twine, and likewise emit rootlets which adhere to any fitting surface.

Part II.—LEAF-CLIMBERS.

It has long been observed that several plants climb by the aid of their leaves, either by the petiole or by the produced midrib; but beyond this simple fact nothing is known of them. Palm and Mohl class these plants with those which bear tendrils; but as a leaf is generally a defined object, the present classification has, at least, some plain advantages. There are other advantages, as leaf-climbers are intermediate in many respects between twiners and certain tendril-bearing plants. I have observed eight species of *Clematis* and seven of *Tropæolum* in order to discover what amount of difference there may be within the same genus; and the differences, as we shall see, are considerable.

CLEMATIS.—*C. glandulosa.*—The thin upper internodes revolve, moving against the course of the sun, precisely like those of a true twiner, at an average rate, judging from three revolutions, of 3 h. 48 m. The leading shoot immediately twined round a stick placed near it; but, after making an open spire of only one turn and a half, it ascended for a short space straight, and then reversed its spire and wound two turns in an opposite course. This was rendered possible by the straight piece between the opposed spires having become rigid. The simple, broad, ovate leaves of this tropical species, so unlike those of most of the other species of the genus, with their short thick petioles, seem but ill-fitted for any movement. Whilst twining up a vertical stick, no use is made of them. Nevertheless, if the footstalk of a young leaf be rubbed with a thin twig a few times on any side, it will in the course of a few hours bend to that side; afterwards it becomes straight again. The under side seemed to be the most sensitive; but the sensitiveness or irritability is but slight com-

pared to that which we shall meet with in some of the following species; for a loop of string, weighing 1·64 grain, hanging for some days on a young footstalk, produced a scarcely perceptible effect. A sketch is here given of two young leaves which had naturally caught two twigs on each side of the stem. A forked twig placed so as to lightly press on the under side of a young footstalk caused it, in 12 h., to bend greatly, and ultimately to such an extent that the leaf passed to the opposite side of the stem;

Fig. 1.

Clematis glandulosa, with two young leaves clasping twigs, with the clasping portions thickened.

the forked stick having been removed, the leaf slowly recovered its proper position.

The young leaves change their position in a rather odd manner: when first developed the petioles are upturned, parallel to the stem; they then slowly bend downwards, remaining for a short time at right angles to the stem, and then become so much arched downwards that the blade of the leaf points to the ground with its tip curled inwards, so that the whole petiole and leaf together form a hook. If they come into contact with no object, they retain this position for a considerable time, and then bending upwards they reassume their original upturned position, which is retained ever afterwards. The young leaves, being hooked, are thus enabled to catch twigs when brought into contact with them by the revolving movement of the internodes. The petioles which have clasped any object soon become much thickened and strengthened, as may be seen in the diagram.

Clematis montana.—The long and thin petioles of the leaves, whilst young, are sensitive, and when lightly rubbed bend to the rubbed side, subsequently becoming straight. They are far more sensitive than the petioles of *C. glandulosa*; for a loop of thread weighing a quarter of a grain caused them to bend; a loop weighing only one-eighth of a grain sometimes acted and sometimes did not act. The sensitiveness extends to the angle between the stem and leaf-stalk. I may here state that I ascertained the weights of the string and thread used in all cases by carefully weighing 50 inches in a chemical balance, and then cutting off measured lengths *. The main petiole carries three leaflets; but the short petioles of these leaflets are not sensitive. A young inclined shoot (the plant being in the greenhouse) made a large circle opposed to the course of the sun in 4 h. 20 m., but the next day, being very cold, the time was 5 h. 10 m. A stick placed near the revolving stem was soon struck by the petioles which stand out at right angles, and the revolving movement was arrested. The petiole then began, being excited by the contact, to slowly wind round the stick. When the stick was thin, the petiole sometimes wound twice round it. The opposite leaf was in no way affected. The attitude assumed by the stem after the petiole has clasped a stick, is that of a man standing by a column, who throws his whole arm horizontally round it. With respect to the stem's power of twining, some remarks will be made under *C. calycina.*

Clematis Sieboldi.—A shoot made three revolutions against the sun at an average rate of 3 h. 11 m. The power of twining is like that of the last species. Its leaves are nearly similar, except that

* Our English grain equals nearly 65 milligrammes.

the petioles of the lateral and terminal leaflets are sensitive. A loop of thread, weighing one-eighth of a grain, acted on the main petiole; but it took between two and three days to produce any effect. The leaves have the remarkable habit and power of spontaneously revolving, generally in vertical ellipses, in the same manner, but in a less degree, as will be described under *C. microphylla.*

Clematis calycina.—The young shoots are thin and flexible; one revolved, describing a broad oval, in 5 h. 30 m., and another in 6 h. 12 m.: they followed the course of the sun; but in all the species of the genus the course followed, if observed long enough, would no doubt be found to differ. This is a rather better twiner than the two last species: the stem, when a thin upright stick free from twigs was placed near, sometimes made two spiral turns round it; then, being arrested by the clasping of the petioles, it would run up for a space straight and then generally reversed its course and took one or two spiral turns in an opposite direction. This reversal of the spire occurred in all the foregoing species. The leaves are so small compared with those of most of the other species that the petioles at first seem ill-fitted for clasping. Nevertheless the main service of the revolving movement is to bring them into contact with surrounding objects, which are slowly but securely seized. The young petioles, which alone are sensitive, have their ends bowed a little downwards, so as to be in a slight degree hooked; ultimately the whole leaf becomes flat. I gently rubbed with a thin twig the lower surfaces of two young petioles; and in 2 h. 30 m. they were slightly curved downwards; in 5 h., after being rubbed, the end of one was bent completely back parallel to the basal portion; and in 4 h. subsequently it became nearly straight again. To show how sensitive the young petioles are, I may mention that I put, in order to mark them, short streaks of water-colour on their under sides; an infinitely thin crust was thus formed, but it sufficed in 24 h. to cause both to bend downwards. Whilst the plant is young, each leaf consists of three divided leaflets, which have barely distinct petioles, and these are not then sensitive; but when the plant is well grown, the two lateral and terminal leaflets have long petioles, and these now become sensitive and are capable of clasping in any direction any object.

When the petiole has clasped a twig, it undergoes some remarkable changes, which occur with the several other species, but in a less strongly marked manner, and will be here described once for all. The clasped petiole in the course of two or three days swells greatly, and ultimately becomes nearly twice as thick as the opposite leaf-stalk which has clasped nothing. When thin transverse slices of the two are placed under the microscope their difference is conspicuous: the side of the footstalk which has been in contact with the support is formed of a layer of colourless cells with their longer axes directed from the centre of the petiole, and very much larger than any cells found in the opposite or unchanged petiole; the central cells, also, are in some degree enlarged, and the whole is much indurated. The exterior surface generally becomes bright red. But a far greater change takes place in the nature of the tissues than that which is externally visible: the petiole of the unclasped leaf is flexible, and can be easily snapped, whereas the clasped footstalk acquires an extraordinary toughness and rigidity, so that considerable force is required to pull it into pieces. With this change, great durability is probably acquired; at least this is the case with the clasped petioles of *Clematis vitalba*. The meaning of these changes is plain, namely, that the petioles may firmly and durably support the stem.

Clematis microphylla, var. *leptophylla*.—The long and thin internodes of this Australian species revolve sometimes in one direction and sometimes in an opposite one, describing long, narrow, irregular ellipses or large circles: four revolutions were completed within five minutes of the same average rate of 1 h. 51 m.; so that this species moves more quickly than any other of the genus. The shoots, when placed near a vertical stick, either twine round it or clasp it with the basal portions of their petioles. The leaves whilst young are nearly of the same general shape, and act in the same manner like a hook, as will be described under *C. viticella*; but the leaflets are more divided, as in *C. calycina*, and each segment whilst young terminates in a hardish point, and is much curved downwards and inwards; so that the whole leaf readily catches and becomes entangled with any neighbouring object. The petioles of the young terminal leaflets are acted on by loops of thread weighing $\frac{1}{8}$th and $\frac{1}{16}$th of a grain: the basal portion of the main petiole is much less sensitive, but will clasp a stick against which it presses.

The whole leaf, whilst young, is in continual, spontaneous, slow movement. The stem was secured close to the base of the leaves, and, a bell-glass being placed over the shoot, the movements of the leaves were traced on it during several days. A very irre-

gular line was generally formed; but one day, in the course of
eight hours and three quarters, the figure traced, clearly represented three and a half irregular ellipses, the most perfect one of
which was completed in 2 h. 35 m. The two opposite leaves moved
quite independently of each other. This movement would aid
that of the internodes in bringing the petioles into contact with
surrounding objects. I discovered this spontaneous movement
too late to be enabled to observe the leaves in all the other species; but from analogy I can hardly doubt that the leaves of at
least *C. viticella*, *C. flammula*, and *C. vitalba* move spontaneously;
and, judging from *C. Sieboldi*, this probably is the case with *C.
montana* and *C. calycina*. I ascertained that the simple leaves of
C. glandulosa exhibited no spontaneous revolving movement.

Clematis viticella, var. *venosa*.—In this and the two following
species the power of spirally twining is completely lost, and this
seems due to the lessened flexibility of the internodes and to the
interference caused by the large size of the leaves. But the
revolving movement, though restricted, is not lost. In our present species a young internode, placed in front of a window, made
three narrow ellipses, transversely to the light, at an average rate
of 2 h. 40 m.; when placed so that the movement was to and from
the light, the rate was greatly accelerated and retarded, as in the
case of twining plants. The ellipses were small; the longer diameter, described by the apex of a shoot bearing a pair of not expanded leaves, being only 4⅝ inches, and that by the apex of the
penultimate internode only 1⅛ inch; at the most favourable
period of growth each leaf would hardly be carried to and fro by
the movement of the internodes more than two or three inches,
but, as above stated, it is probable that the leaves themselves
move spontaneously. The movement of the whole shoot by the
wind and by its rapid growth would probably be almost equally
efficient with the spontaneous movements in bringing the petioles
into contact with surrounding objects.

The leaves are of large size. There are three pairs of lateral
leaflets and a terminal one, all borne by rather long petioles. The
main petiole bends a little angularly downwards at each point
where a pair of leaflets arises, and the petiole of the terminal
leaflet is bent downwards at right angles; hence the whole petiole,
with its rectangularly bent extremity, acts as a hook. This,
with the lateral petioles directed a little upwards, forms an excellent grappling apparatus by which the leaves readily become
entangled with surrounding objects. If they catch nothing, the

whole petiole ultimately grows straight. Both the medial and lateral petioles are sensitive; and the three branches, into which the basi-lateral petioles are generally subdivided, likewise are sensitive. The basal portion of the main petiole between the stem and the first pair of leaflets is less sensitive than the remainder, but it will clasp a stick when in contact. On the other hand, the

Fig. 2.

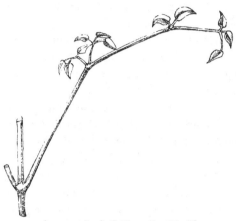

A young leaf of *Clematis viticella.*

inferior surface of the rectangularly bent terminal portion (carrying the terminal leaflet), which forms the inner side of the end of the hook, is the most sensitive part; and this portion is manifestly best adapted to catch distant supports. To show the difference in sensibility, I gently placed loops of string of the same weight (in one instance weighing ·82 of a grain) on the several lateral and on the terminal sub-petioles; in a few hours the latter were bent, but after 24 h. no effect was produced on any of the lateral petioles. Again, a terminal sub-petiole placed in contact with a thin stick became sensibly curved in 45 m., and in 1 h. 10 m. had moved through ninety degrees, whereas a lateral petiole did not become sensibly curved until 3 h. 30 m. had elapsed. In this latter case, and in all other such cases, if the sticks be taken away, the petioles continue to move during many hours afterwards; so they do after a slight rubbing; but ultimately, if the flexure has not been very great or long-continued, they become, after about a day's interval, straight again.

The gradation in the extension of the sensitiveness in the petioles of the several above-described species deserves notice. In *C. montana* it is confined to the main petiole, and has not spread to the sub-petioles of the three leaflets; so it is with young plants of *C. calycina*; but in older plants it has spread to the three sub-petioles. In *C. viticella* it has spread to the petioles of the seven leaflets, and to the subdivisions of the basi-lateral sub-petioles. In this latter species the sensitiveness has diminished in the basal part of the main petiole, in which alone it resided in *C. montana*, and has accumulated in the abruptly bent terminal portion.

Clematis flammula.—The shoots, which are rather thick, straight, and stiff, whilst growing vigorously in the spring, made small oval revolutions, following the sun in their course. Four were made at an average rate of 3 h. 45 m. The longer axis of the oval, described by the extreme tip, was directed at right angles to the line joining the opposite leaves; its length was in one case only $1\frac{3}{8}$, and in another case $1\frac{6}{8}$ inch; so that the young leaves are moved a very short distance. The shoots of the same plant observed in midsummer, when growing not so quickly, did not revolve at all. I cut down another plant in the early summer, so that by August 1st it had formed new and moderately vigorous shoots; these, when observed under a bell-glass, were on some days quite stationary, and on other days moved to and fro only about the eighth of an inch. Consequently the revolving power is here much enfeebled, and under unfavourable circumstances is completely lost. This species must depend on the probable, though not ascertained, spontaneous movements of its leaves, on the rapid growth of its shoots, and on movements from the wind, for coming into contact with surrounding objects: hence, perhaps, it is that the petioles have acquired, as we shall see, in compensation a high degree of sensitiveness.

The petioles are bowed downwards, and have the same general hook-like form as in *C. viticella*. The medial petiole and lateral sub-petioles are sensitive, especially the much-bent terminal portion. As the sensitiveness is here greater than in any other species of the genus observed by me, and is in itself remarkable, I will give fuller details. The petioles, when so young that they have not separated from each other, are not sensitive; when the lamina of a leaflet has grown to quarter of an inch in length (that is, about one-sixth of its full size), the sensitiveness is highest; but at this period the petioles are much more fully developed proportionally than the laminæ of the leaves. Full-grown petioles are not in the least sensitive. A thin stick placed so as to press lightly against a petiole, bearing a leaflet a quarter of an inch in length, caused the petiole to bend in 3 h. 15 m.; in another case a petiole curled completely round a stick in 12 h. These petioles were left curled for 24 h., and then the sticks were removed; but they never straightened themselves. I took a twig, thinner than the petiole itself, and lightly rubbed with it several petioles four times up and down; these in 1 h. 45 m. became slightly curled; the curvature increased during some hours and then began to decrease, but after 25 h. from the time of rubbing a vestige of the curvature remained. Some other petioles similarly rubbed once up and down became perceptibly curved in about 2 h. 30 m., a terminal sub-petiole moving more than a lateral sub-petiole; they became quite straight again in between 12 h. and 14 h. Lastly, a length of about one-eighth of an inch of a sub-petiole, lightly rubbed with the same twig only once down, became slightly curved in 3 h., and remained so during 11 h., but the next morning was quite straight.

The following observations are more precise. After finding that heavier pieces of string and thread acted, I placed a loop of string, weighing 1·04 gr., on a terminal petiole: in 6 h. 40 m. a curvature could be seen; in 24 h. the petiole formed an open ring round the string; in 48 h. the ring had almost closed on the string, and in 72 h. it had firmly seized the fine twine so that it required some force to withdraw it. A loop weighing ·52 of a grain caused a *lateral* sub-petiole just perceptibly to curve in 14 h., but after 24 h. it had moved through ninety degrees. These observations were made during the summer: the following were made in the spring, when the petioles are apparently more sensitive:—A loop of thread, weighing one-eighth of a grain, produced no effect on the lateral sub-petioles, but placed on a terminal one caused, after 24 h., a moderate curvature in it; the curvature, though the loop remained suspended, was after 48 h. diminished, but never disappeared, showing that the petiole had become partially accustomed to the insufficient stimulus. This experiment was twice repeated with nearly similar results. Lastly, a loop of thread, weighing only one-sixteenth of a grain (nearly equal to four milligrammes), was twice gently placed by a forceps on a terminal sub-petiole (the plant being, of course, in a still and closed room), and this weight certainly caused a flexure, which very slowly increased until the petiole had moved through nearly ninety degrees: beyond this it did not move; nor did the petiole, the loop remaining suspended, ever become perfectly straight again.

When we consider, on the one hand, the thickness and stiffness of the petioles, and, on the other hand, the thinness and softness of fine cotton thread, and what an extremely small weight one-

D

sixteenth of a grain is, these facts are remarkable. But I have
reason to believe that even a less weight causes a curvature when
acting over a broader surface than can be affected by thin thread.
Having noticed that the tail of a suspended string, which accidentally touched a petiole, had caused it to bend, I took two
pieces of thin twine, 10 inches in length (weighing 1·64 gr.), and,
tying them to a stick, let them hang as nearly perpendicularly
downwards as their thinness and flexuous nature, after being
stretched, would permit; I then quietly placed their ends so as
just to rest on two petioles with their tips hanging about the
tenth of an inch beneath; both these petioles certainly became
curved in 36 h. One of the ends of string, which just touched
the angle between a terminal and lateral sub-petiole, was in 48 h.
caught as by a forceps between them. In these cases the pressure,
though spread over a wider surface than that touched by the
cotton thread, must have been excessively slight.

Clematis vitalba.—My plants in pots were not healthy; so that
I dare not trust my observations, which indicated much similarity
in habits with *C. flammula.* I mention this species only because
I saw many proofs that the petioles of plants growing naturally
are excited to movement by very slight pressure. For instance,
I found petioles which had clasped thin withered blades of grass,
the soft young leaves of a maple, and the lateral flower-peduncles
of the quaking-grass or Briza: the latter are only about as thick
as a hair from a man's beard, but they were completely surrounded
and clasped. The petioles of a leaf, so young that none of the
leaflets had expanded, had partially seized on a twig. The petioles
of almost every old leaf, even when unattached to any object, are
much convoluted; but this is owing to their having come, whilst
young, into contact during several hours with some object subsequently removed. With the several above-described species,
cultivated in pots and thus carefully observed, there never was
any bending of the petioles without the stimulus of contact. When
winter comes on, the blades of the leaves of *C. vitalba* drop off;
but the petioles (as was also observed by Mohl) remain, sometimes during two seasons, attached to the branches; and, being
convoluted, they curiously resemble true tendrils, such as those
occurring in the allied genus *Naravelia.* The petioles which
have clasped an object become much more woody, stiff, hard,
and polished than those which have failed in this their proper
purpose.

TROPÆOLUM.—I observed *T. tricolorum, T. azureum, T. penta-*

phyllum, T. peregrinum, T. elegans, T. tuberosum, and a dwarf variety of, as I believe, *T. minus.*

Tropæolum tricolorum, var. *grandiflorum.*—The flexible shoot, which first rises from the tuber, is as thin as thin twine. One such shoot revolved in a course opposed to the sun, at an average rate, judging from three revolutions, of 1 h. 23 m.; but no doubt the direction of the revolving movement is variable. When the plant had grown tall and much branched, all the many lateral shoots continued to revolve. The stem, whilst young, twined regularly round a thin vertical stick; in one case I counted eight spiral turns: but when grown older, the stem often runs straight up for a space, and, being arrested by the clasping petioles, makes one or two spires in a reversed direction. Until the plant has grown to a height of two or three feet, about a month after the first shoot has appeared above ground, no true leaves, but in their place little filaments, coloured like the stem, are produced. The extremities of these filaments are pointed, a little flattened, and furrowed on the upper surface. They never become developed into leaves. As the plant grows in height new filaments are produced with slightly enlarged tips; then others, bearing on each side of the enlarged medial tip a rudimentary segment of a leaf; and soon other segments appear, until a perfect leaf is formed with seven deep segments. So that on the same plant we may see every step from tendril-like filaments to perfect leaves. Hence this plant, whilst young, might be classed with tendril-bearers. After the plant has grown to a considerable height, and is secured to its support by the clasping petioles of the true leaves, the clasping filaments on the lower part of the stem wither and drop off; so that they perform only a temporary service.

These filaments, as well as the petioles of the perfect leaves, whilst young, are highly sensitive on all sides to a touch. The slightest rub causes them to curve towards the rubbed side in about three minutes: one bent itself into a ring in six minutes; they subsequently became straight again: if, however, they have once completely clasped a stick, when this is removed, they do not recover themselves. The most remarkable fact, and which I have observed in no other species of the genus, is that the filaments and petioles of the young leaves, if they catch no object, after standing in their original position for some days spontaneously and slowly move, oscillating a little from side to side, towards the stem of the plant. Hence all the petioles and filaments, though arising on different sides of the axis, ultimately bend towards and clasp either their own stem or the supporting stick. The petioles and filaments often become, after a time, in some degree spirally contracted. In these spontaneous movements, and in the abortion of their laminæ, the sensitive filaments present a much nearer approach to the condition of tendrils than do the petioles of any other leaf-climber observed by me.

Tropæolum azureum.—An upper internode made four revolutions, following the sun, at an average rate of 1 h. 47 m. The stem twined spirally in the same irregular manner as in the last species; it produced no filaments or rudimentary leaves. The petioles of the young leaves are very sensitive: a single very light rub with a twig caused one to move perceptibly in 5 m., and another in 6 m.; the former petiole became bent at right angles in 15 m., and became straight again in between 5 h. and 6 h. A loop of thread weighing ⅓th of a grain caused a petiole to curve.

Tropæolum pentaphyllum.—The plant observed by me had not the power of spirally twining, which seemed due, not to the want of flexibility in the stem, but rather to continual interference from the clasping petioles. An upper internode made three revolutions, following the sun, at an average rate of 1 h. 46 m. The main purpose of the revolving movement in all the species is manifestly to bring the petioles into contact with some supporting object. The petiole of a young leaf, after a slight rub, became curved in 6 m.; another, on a cold day, in 20 m.; but others generally in from 8 m. to 10 m.: the curvature usually increased greatly in from 15 m. to 20 m. The petioles became straight again in between 5 h. and 6 h., and on one occasion in 3 h. When a petiole had fairly clasped a stick, it could not on the removal of the stick recover itself; but the free upper part of a petiole, which had already clasped a stick by its basal part, still had the power of movement. A loop of thread weighing ⅓th of a grain certainly caused a petiole to curve; but the stimulus was not sufficient, the loop remaining suspended, to cause a permanent flexure. If a much heavier loop be placed in the angle between the petiole and the stem, it produces no effect; whereas we have seen that the angle between the stem and petiole of *Clematis montana* is sensitive.

Tropæolum peregrinum.—In a very young plant the internodes did not revolve, resembling in this respect a young twining plant. The four upper internodes in an older plant made three irregular revolutions, in a course opposed to the sun, at an average rate of 1 h. 48 m. It is remarkable how nearly the same the average rate of revolution (taken, however, but from few observations) is in this and the two last species, namely, 1 h. 47 m., 1 h. 46 m, and 1 h. 48 m. The present species cannot spirally twine, which seems mainly due to the rigidity of its stem. In a very young plant, which did not revolve, the petioles were not sensitive. In older plants the petioles of quite young leaves, and of leaves as much as an inch and a quarter in diameter, are sensitive. A moderate rub caused one to curve in 10 m., but others in 20 m.; the petioles became straight again in from 5 h. 45 m. to 8 h. Petioles which have naturally come into contact with a stick, sometimes take two turns round it. When clasped round a support, they become rigid and hard. The petioles are less sensitive to a weight than in the previous species; for loops of string weighing ·82 of a grain did not cause any curvature, whilst a loop of double this weight (1·64 gr.) did act.

Tropæolum elegans.—I did not make many observations on this species. The short and stiff internodes revolve irregularly, and describe extremely small oval figures; one was completed in 3 h. A young petiole, when rubbed, became slightly curved in 17 m.; then much more so; and was nearly straight again in 8 h.

Tropæolum tuberosum.—The internodes on a plant nine inches high did not move at all; but on an older plant they moved irregularly, and made very small imperfect ovals. These movements could be detected only by being traced on a bell-glass placed over the plant. Sometimes the shoots stood still for hours; during some days they moved only in one direction in a crooked line; on other days they made small irregular spires or circles, one being completed in about 4 h. The movement of the apex of the shoot, from extreme point to point of the oval, was only about one inch or one and a half; yet this slight movement brought the petioles into contact with closely surrounding twigs, which were then clasped. With the lessened power of spontaneously revolving, compared with the previous species, the sensitiveness of the petioles is likewise diminished. These, when rubbed a few times, did not become curved until half an hour had elapsed; the curvature increased during the next two hours, and then very slowly decreased; so that the petioles sometimes required 24 h. to become straight again. The petioles of very young leaves can act perfectly; one with the lamina only ·15 of an inch in diameter, that is, about a twentieth of the full size, firmly clasped a thin twig: but leaves grown to one quarter of their full size can likewise act.

Tropæolum minus (?).—The internodes of a variety named

" dwarf crimson Nasturtium " had no power of revolving ; but they moved during the day to the light, and from it at night, in a rather irregular course. The petioles, when well rubbed, showed no power of curving ; nor could I see that they ever clasped any neighbouring support. We have seen in this genus a gradation from species such as *T. tricolorum*, which have exquisitely sensitive petioles, and internodes which have rapid revolving powers and can spirally twine up a support, to other species, such as *T. elegans* and *T. tuberosum*, the petioles of which are much less sensitive, and the internodes of which have very feeble revolving powers and cannot spirally twine round a support, to this last species, which has entirely lost or never acquired these faculties. From the general character of the genus, the loss of power seems the more probable alternative.

In this species and in *T. elegans*, and probably in others, the flower-peduncles, as soon as the seed-capsule begins to swell, spontaneously bend abruptly downwards and become somewhat convoluted : when a stick lies in the path, it is to a certain extent clasped ; but, as far as I have been able to observe, the movement of the peduncle is quite independent of the stimulus from contact.

ANTIRRHINEÆ.—In this tribe (Lindley) of the Scrophulariaceæ, at least four of the seven included genera have leaf-climbing species.

Maurandia Barclayana.—A thin, slightly bowed shoot made two revolutions, following the sun, each in 3 h. 17 m. ; this same shoot, the day before, revolved in an opposite direction. The shoots do not spirally twine, but climb excellently by the aid of the young sensitive petioles. These petioles, when lightly rubbed, move after a considerable interval of time, and subsequently become straight again ; a loop of thread weighing $\frac{1}{8}$th of a grain caused them to bend.

Maurandia semperflorens.—This freely growing species climbs exactly like the last, by its sensitive petioles. A young internode made two circles, each in 1 h. 46 m. ; so that it moves almost twice as rapidly as the last species. But I should not have noticed the present species, had it not been for the following unique case. Mohl says (S. 45) that "the flower-peduncles, as well as the petioles, are wound into tendrils ;" and he adds nothing more about the genus. But it must be observed that Mohl classes as tendrils even such objects as the spiral flower-stalks of the *Vallisneria*. Nevertheless this remark, and the well-known fact that the flower-peduncles of this *Maurandia* are flexuous, led me carefully to examine them. They never act as tendrils : I repeatedly placed thin sticks in contact with young and old peduncles, and I allowed nine vigorous plants to grow over an entangled mass of branches ; but in no one instance did a peduncle bend round any object. It is indeed in the highest degree improbable that this should occur, for the flower-peduncles are generally developed on branches which have already securely clasped a support by their petioles ; and when borne on free depending branches, they are not produced by the terminal portion of the internode which alone has the power of revolving; so that they can only accidentally and rarely be brought into contact with any surrounding object. Nevertheless (and this is the remarkable fact) these flower-peduncles, whilst young, exhibit feeble revolving powers, and are slightly sensitive to a touch. I selected some stems which had firmly clasped a stick by their petioles, and, placing a bell-glass over them, traced the movements of the young flower-peduncles. Some days these moved over a short and extremely irregular line, making little loops in their course. One day a young peduncle 1½ inch in extreme length was carefully observed, and it made four and a half narrow, vertical, irregular, and very short ellipses—each at an average rate of about 2 h. 25 m. ; an adjoining peduncle described during the same time similar, but fewer, ellipses. As the plant had for some time occupied exactly the same position, these movements could not be attributed to the varying action of the light. Peduncles, old enough for the coloured petals to be just visible, do not move. With respect to irritability, I rubbed a few times very lightly with a thin twig two young peduncles (1½ inch in length), one on the upper side and the other on the lower side, and they became in between 4 h. and 5 h. plainly bowed towards the rubbed sides ; in 24 h. subsequently, they straightened themselves. Next day they were rubbed on the opposite sides, and they became perceptibly curved towards these sides. Two other and younger peduncles (three-fourths of an inch in length) were lightly rubbed on their adjoining sides, and they became so much bowed towards each other, that the arcs of the bows stood at nearly right angles to their previous positions ; this was the greatest movement seen by me ; subsequently they straightened themselves. Other peduncles, so young as to be only three-tenths of an inch in length, became curved when rubbed. On the other hand, peduncles above 1½ inch in length required to be rubbed two or three times, and then became only just perceptibly curved. Loops of thread suspended on the peduncles produced no effect; but loops of string weighing ·82 and 1·64 grain acted capriciously, sometimes causing a slight curvature; but they were never clasped, like the far lighter loops of thread by the petioles.

In the nine vigorous plants which I observed, it is certain that neither the slight spontaneous movements nor the slight sensitiveness of the flower-peduncles were of any service to the plants in climbing. If any member of the Scrophulariaceæ had been known to have flower-peduncles used for climbing, or had tendrils produced by their modification, I should have thought that this *Maurandia* still retained a useless or rudimentary vestige of a former habit; but this view cannot be maintained. We are almost compelled to believe that by some correlation of growth the power of movement has been transferred from the young internodes to the young peduncles, and in the same manner sensitiveness from the young petioles to the young peduncles; but this latter supposition is the more improbable, as I could detect no sensitiveness in the young internodes of the *Maurandia*, though in a closely allied genus, *Lophospermum*, the young internodes, as we shall see, are sensitive. By whatever means the peduncles of this *Maurandia* have acquired their power of spontaneous movement and their sensitiveness, the case is interesting for us; for we can see that if these now useless capacities were a little perfected, the flower-peduncles could be made as useful for climbing as are the flower-peduncles of *Vitis* and *Cardiospermum*, as will hereafter be described.

Rhodochiton volubile.—A long flexible shoot swept a large circle, following the sun, in 5 h. 30 m.; and, as the day became warmer, a second circle in 4 h. 10 m. The shoots sometimes make a whole or half spire round a vertical stick, then run up for a space straight, and afterwards make spiral turns in an opposite direction. The petioles of very young leaves, about one-tenth of their full size, are highly sensitive, and bend towards any side which has been touched; but they do not move quickly : one, after being lightly rubbed, was perceptibly curved in 1 h. 10 m., and became considerably arched in 5 h. 40 m. after the rubbing; some other petioles, after being rubbed, were scarcely curved in 5 h. 30 m., but in 6 h. 30 m. were distinctly curved. A curvature was perceptible in a petiole in between 4 h. 30 m. and 5 h., after the suspension of a little loop of string. A loop of fine cotton thread, weighing one-sixteenth of a grain, not only slowly caused a petiole to bend, but was ultimately firmly clasped by it, so that it could be withdrawn only by some little force. The petioles, when coming into contact with a stick, take either a complete or half turn round it ; ultimately they increase much in thickness. Leaves arising on the side of the stem opposite to the light move towards it ; and, in doing so, the petioles are sometimes brought into contact with the stem, and consequently clasp it ; but the petioles have no true spontaneous movement.

Lophospermum scandens, var. *purpureum.*—Some long, moderately thin internodes made four revolutions at an average rate of 3 h. 15 m. The course pursued was very irregular—sometimes an extremely narrow ellipse, sometimes a large circle, sometimes an irregular spire or zigzag line, and sometimes the apex stood still. The young petioles, when brought by the revolving movement into contact with a stick, clasp it, and soon increase considerably in thickness ; but they are not quite so sensitive to a light weight as those of the *Rhodochiton*, for loops of thread weighing one-eighth of a grain did not invariably cause them to bend.

This plant presents a case not observed in any other leaf-climber or twiner or tendril-bearer, or in any other plant as far as I know, namely, that the young internodes are sensitive to a touch. When a petiole clasps a stick, it draws the base of the internode against it ; and then the internode itself bends towards the stick, which is thus caught between the stem and the petiole as by a pair of pincers. The internode straightens itself again, excepting the part in contact with the stick. Young internodes alone are sensitive, and these are sensitive on all sides along their whole length. I made fifteen trials by lightly rubbing two or three times with a thin twig several internodes ; and in about 2 h., but in one case in 3 h., all became bent : they became straight again in about 4 h., subsequently. An internode, which was rubbed as much as six or seven times with a twig, became just perceptibly curved in 1 h. 15 m., and subsequently in 3 h. the curvature increased much ; the internode became straight again in the course of the night. I rubbed some internodes one day on one side, and the next day on the opposite side or at right angles ; and the curvature was always towards the rubbed side.

According to Palm (S. 63), the petioles of *Linaria cirrhosa* and, to a limited degree, those of *L. elatine* have the power of clasping a support.

SOLANUM.—*S. jasminoides.*—Some of the species of this large genus are twiners ; but this is a true leaf-climber. A long, nearly upright shoot made four revolutions, moving against the sun, very regularly at an average rate of 3 h. 26 m. The shoots, however, sometimes stand still. It is considered a greenhouse plant; but when kept there, the petioles took several days to clasp a stick: in the hothouse a stick was clasped in 7 h. In the greenhouse a

Fig. 3.

Solanum jasminoides, with one of its leaves clasping a stick.

petiole was not affected by a loop of string, suspended during several days and weighing 2½ grains; in the hothouse one was made to curve by a loop weighing 1·64 (and, on the removal of the string, became straight again), but was not at all affected by another loop weighing ·82 of a grain. We have seen that the petioles of some other leaf-climbing plants were affected by one-thirteenth of this latter weight. In this plant, and in no other leaf-climber seen by me, a leaf grown to its full size was capable of clasping a stick; but the movement was so extraordinarily slow that in the greenhouse the act required several weeks; but on each succeeding week it was clear that the petiole became more and more curved, until finally it firmly clasped the stick.

When the flexible petiole of a half- or a quarter-grown leaf has clasped any object, in three or four days it increases much in thickness, and after several weeks becomes wonderfully hard and rigid; so that I could hardly remove one from its support. On comparing a thin transverse slice of this petiole with one from the next or older leaf beneath, which had not clasped anything, its diameter was found to be fully doubled, and its structure greatly changed. In two other petioles similarly compared, and here represented, the increase in diameter was not quite so great. In the section of the petiole in its ordinary state (A), we see a

Fig. 4.

Solanum jasminoides.

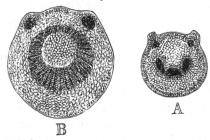

A. Section of a petiole.
B. Section of a petiole some weeks after it had clasped a stick, as shown in fig. 3.

semilunar band of cellular tissue slightly different from that outside it, and including three closely approximate groups of dark vessels. Near the upper surface of the petiole, beneath two ridges, there are two other small circular groups of vessels. In the section of the petiole (B) which had during several weeks clasped a stick, the two upper ridges have become much less prominent, and the two groups of woody vessels beneath them much increased in diameter. The semilunar band is converted into a complete ring of very hard, white, woody tissue, with lines radiating from the centre. The three groups of vessels, which, though closely approximate, were before distinct, are now completely blended together. The upper part of the new ring of woody vessels, formed by the prolongation of the horns of the original semilunar band, is thinner than the lower part, and is slightly different in appearance from being less compact. This clasped petiole had actually become thicker than the stem close beneath; and this was chiefly due to the greater thickness of the ring of wood, which presented, both in transverse and longitudinal sections, a closely similar structure in the petiole and axis. The assumption by a petiole of this structure is a singular morphological fact; but it is a still more singular physiological fact that so great a change should have been induced by the mere act of clasping a support *.

FUMARIACEÆ.—*Fumaria officinalis.*—It could not have been

* Dr. Maxwell Masters informs me that in most, or all, petioles which are cylindrical, such as those bearing peltate leaves, the woody vessels form a closed ring, and that the semilunar band of vessels is confined to petioles which are channelled along their upper surfaces. In accordance with this statement, it

anticipated that so lowly a plant would have been a climber. This
it effects by the aid of the main and lateral petioles of its com-
pound leaves; even the much-flattened terminal portion of the
petiole can seize a support. I have seèn a substance as soft as a
withered blade of grass caught. Petioles which have clasped any
object ultimately became rather thicker and more cylindrical.
On lightly rubbing with a twig several petioles, they became perceptibly curved in 1 h. 15 m., and subsequently they straightened
themselves. A stick gently placed in the angle between two sub-
petioles caused movement in 7 h., and was almost clasped in 9 h. A
loop of thread, weighing one-eighth of a grain, caused, after 12 h.
and before 20 h. had elapsed, a considerable curvature; but the
petiole never fairly clasped the thread. The young internodes
are in continual movement; the movement is considerable, but
very irregular in course; a zigzag line, or a spire crossing itself,
or a figure of 8 is formed; the course during 12 h., being traced
on a bell-glass, apparently represented about four ellipses. The
leaves themselves also move spontaneously, the main petiole curving itself in accordance with the movement of the internodes; so
that when the latter move to one side the petiole is curved to
that side, then, becoming straight, is curved to the opposite side.
Thus a wider space is swept for a support to be clasped. The
movement, however, is small, as could be seen when the shoot
was securely tied to a stick and the leaf alone allowed to move.
The leaf in this case followed an irregular course, like that made
by the young internodes.

Adlumia cirrhosa.—I raised some plants late in the summer;
they formed magnificent leaves, but threw up no central stem.
The first-formed leaves were not sensitive; but some of the later
leaves were sensitive, but only towards their extremities, and were
able to clasp sticks. This could be of no service to the plant, as
these leaves rose from the ground; but it showed what the future
character of the plant would be when it had grown tall enough to
climb. The tip of one of these ground leaves, whilst young, described in 1 h. 36 m. a narrow ellipse, open at one end, and exactly
three inches in length; a second ellipse was broader, more irregular, and shorter, viz. only 2½ inches in length, and was completed in 2 h. 2 m. From analogy with *Fumaria* and *Corydalis*, I
have no doubt that the internodes have the power of revolving.

may be observed that the enlarged and clasped petiole of the *Solanum*, with its
closed ring of woody vessels, has become much more cylindrical than it was in
its original unclasped condition.

Corydalis claviculata.—This plant is interesting from being in a condition so exactly intermediate between a leaf-climber and a tendril-bearer that it might have been described under either head ; but, for reasons hereafter assigned, it is classed amongst tendril-bearers.

Besides the plants already described, *Bignonia unguis* and its close allies, though aided by tendrils, as will hereafter be described, have clasping petioles. According to Mohl (S. 40), *Cocculus Japonicus* (one of the Menispermaceæ) and a fern, the *Ophioglossum Japonicum* (S. 39), climb by their leaf-stalks.

We now come to a small section of plants which climb by the aid of the produced midribs or tips of their leaves.

GLORIOSA.— *G. Plantii* (Liliaceæ).—The stem of a half-grown plant continually moved, generally describing an irregular spire, but sometimes ovals, with the longer axes running in different directions. It either followed the sun, or moved in an opposite course, and sometimes stood still before reversing its course. One oval was completed in 3 h. 40 m. ; of two horseshoe-shaped figures, one was completed in 4 h. 35 m. and the other in 3 h. The tip of the shoot, in its movements, reached points between four and five inches asunder. The young leaves, when first developed, stand up nearly vertically ; but by the growth of the axis, and by the spontaneous bending down of the terminal half of the leaf, they soon become much inclined, and ultimately horizontal. The end of the leaf forms a narrow, ribbon-like, thickened projection, which at first is nearly straight ; but by the time the leaf has got into an inclined position, the end has bent itself downwards into a well-formed hook ; and this is now strong and rigid enough to catch any object, and, when caught, to anchor the plant and stop the revolving movement. This hook is sensitive on its inner surface, but not in nearly so high a degree as with the many before-described petioles ; for a loop of string, weighing 1·64 grain, produced no effect. When the hook has caught a thin twig or even a rigid fibre, the point may be perceived in from 1 h. to 3 h. to have curled a little inwards ; and, under favourable circumstances, in from 8 h. to 10 h. it finally curls round and seizes the object, which it never again looses. The hook when first formed, before the leaf has become inclined, is less sensitive. The hook, if it catches hold of nothing, remains for a long period open and sensitive ; ultimately the tip spontaneously and slowly curls inwards, and makes a button-like, flat, spiral coil at the end of the leaf. One leaf was watched, and the hook remained open for thirty-three days; but during the last week the tip had curled inwards so much that at last only a very thin twig could have been inserted. As soon as the curling-in of the tip has closed the hook and converted it into a ring, its sensibility, both within and without, is lost; but as long as the hook remains open its sensibility is retained.

When the plant had grown from the bulb to the height of only about six inches, the leaves, four or five in number, were broader than those subsequently produced, and their soft and but little-attenuated tips did not form hooks, and were not sensitive; nor did the stem revolve. At this early period of growth, the plant can support itself; its climbing apparatus is not required, and therefore is not acquired. On the other hand, a full-grown plant which was flowering, and which would not have grown any taller, had leaves on the summit, which were not sensitive, and could not clasp a stick.

Flagellaria Indica (Commelynaceæ).—From dried specimens it is manifest that this plant climbs exactly like *Gloriosa*. A young plant, 12 inches in height, and bearing fifteen leaves, had not one leaf as yet produced into a hook or tendril-like filament; nor did the stem revolve. Hence this plant acquires its climbing power later in life than the *Gloriosa* lily. According to Mohl (S. 41), *Uvularia* (Melanthaceæ) climbs like *Gloriosa*.

These three last-named genera are all Monocotyledons; but there is one Dicotyledon, namely *Nepenthes*, which is ranked by Mohl (S. 41) amongst tendril-bearers; and I hear from Dr. Hooker that most of the species climb well at Kew. This is effected by the stalk or midrib between the leaf and the pitcher twisting round any support. The twisted part becomes thicker; but I observed at Mr. Veitch's that the stalk often takes a turn when not in contact with any object, and that this twisted part likewise becomes thickened. Two vigorous young plants of *N. lævis* and *N. distillatoria*, in my hothouse, whilst less than a foot in height, showed no sensitiveness in their leaves or power of movement or of climbing. But when *N. lævis* had grown to a height of 16 inches, there were signs of these powers. Each young leaf when first formed stands upright, but soon becomes inclined; at this period of growth it terminates in a stalk or filament, with the pitcher at the extremity so little developed that this part is not thicker than any other part. The leaf in this state certainly exhibited slight spontaneous movements; and when the stalk came into contact with a stick, it very slowly bent round and firmly seized it. But the leaf by its subsequent growth became quite slack, though the terminal stalk remained coiled round the stick ; hence it would appear that the chief use of the coiling, at least whilst the plant is young, is to support the pitcher with its load of secreted fluid.

Summary on Leaf-climbers.—Plants belonging to eight families are known to have clasping petioles, and plants belonging to four families climb by the tips of their leaves. With all the plants observed by me, the young internodes revolved more or less regularly, in some cases as regularly as does any twining plant, and at various rates, but generally rather rapidly. Some few can ascend by twining spirally round a support. Differently from most twiners, there is a strong tendency in the same shoot to revolve first in one and then in the opposite direction. The object gained by the revolving movement, as could be plainly seen, was to bring the petioles or the tips of the leaves into contact with surrounding objects ; without this aid there would be a poor chance of success. With rare exceptions, the petioles are sensitive only whilst young ; they are sensitive on all sides, but in different degrees in different plants, and in some species of *Clematis* in very different degrees in different parts of the same petiole. The hooked tips of the leaves of the *Gloriosa* are sensitive only on their inner or inferior surface. The petioles are sensitive to a touch and to excessively slight continued pressure, even from a loop of soft thread weighing only the one-sixteenth of a grain ; and there is reason to believe that the rather thick and stiff petioles of *Clematis flammula* are sensitive to even a less weight when spread over a wider surface. The petioles always bend towards the touched or pressed side, at different rates in different plants, sometimes within a few minutes, but generally after a much longer period. After temporary contact with any object, the petiole continues to bend for a considerable time ; afterwards it slowly becomes straight again, and can then re-act. A petiole excited by an extremely slight weight sometimes bends a little, and then becomes habituated to the stimulus, and either bends no more or becomes straight again, the weight still remaining suspended. Petioles which have clasped any object for some little time cannot recover their original position. After remaining clasped for two or three days, they generally increase much in thickness, either throughout or on one side alone ; they subsequently become, sometimes in a wonderful degree, stronger and more woody ; and in some cases they acquire an internal structure like that of the stem or axis.

The young internodes of the *Lophospermum* are sensitive as well as the petioles, and by their combined movement seize any object. The flower-peduncles of the *Maurandia semperflorens* revolve spontaneously, and are sensitive to a touch, yet are certainly useless for climbing. The leaves of at least two and probably of most of the species of *Clematis*, and of *Fumaria* and *Adlumia*, spontaneously curve from side to side, like the internodes, and are thus better adapted to seize any distant object. The petioles of the perfect leaves, as well as the rudimentary or tendril-like leaves of *Tropæolum tricolorum* move spontaneously and slowly towards their own stem or the supporting stick, which they then clasp; these petioles also show some tendency to contract spirally. The tips of the uncaught leaves of the *Gloriosa*, as they grow old, contract into a flat spire. These several facts are interesting, as we shall see, in relation to true tendrils.

It was observed in some cases that, as with twining plants, so with leaf-climbers, the first internodes which rise from the ground do not spontaneously revolve; nor are the petioles or tips of the first-formed leaves sensitive. In certain species of *Clematis* the high development and spontaneous movements of the leaves, with their highly sensitive petioles, apparently have rendered almost superfluous the spontaneous movements of the internodes, which have consequently become enfeebled. In certain species of *Tropæolum* it would appear as if both the spontaneous movements of the internodes and the sensitiveness of the petioles have become enfeebled; and in one species they have been completely lost.

Part III.—Tendril-bearing Plants.

By tendrils I mean filamentary organs, sensitive to contact and used exclusively for climbing. By this definition, spines or hooks and rootlets, all of which are used for climbing, are excluded. True tendrils are formed by the modification of leaves with their petioles, of flower-peduncles, perhaps also of branches and stipules. Mohl, who includes with true tendrils various organs having a similar external appearance, classes them according to their homological nature, as being modified leaves, flower-peduncles, &c. This would be an excellent scheme; but I observe that botanists, who are capable of judging, are by no means unanimous on the nature of certain tendrils. Consequently I will describe tendril-bearing plants by natural families, following Lindley, and this will in most, or in all, cases keep those of the same homological nature together; but I shall treat of each family, one after the other, according to convenience*. The species to be described belong to ten families, and will be given in the following order :—*Bignoniaceæ, Polemoniaceæ, Leguminosæ, Compositæ, Smilaceæ, Fumariaceæ, Cucurbitaceæ, Vitaceæ, Sapindaceæ, Passifloraceæ.*

BIGNONIACEÆ.—This family contains many tendril-bearers, some twiners, and some root-climbers. The tendrils are always modified leaves. Nine species of *Bignonia*, selected by hazard, are here described, in order to show what diversity of structure and action there may be in species of the same genus, and to show how re· markable the action of the tendrils may be in some cases. The species, taken together, afford connecting links between twiners, leaf-climbers, tendril-bearers, and root-climbers.

Bignonia (an unnamed species from Kew, closely allied to *B. unguis*, but with smaller and rather broader leaves).—A young shoot from a cut-down plant made three revolutions against the sun, at an average rate of 2 h. 6 m. The stem is thin and flexible and twined, ascending, from left to right, round a slender vertical stick as perfectly and as regularly as any true twining-plant. When thus ascending, it makes no use of its tendrils or its petioles ; but when it twined round a rather thick stick, and its petioles were brought into contact with it, these curved round the stick, showing that they have some degree of irritability. The petioles also exhibit a slight

Fig. 5†.

Bignonia, unnamed species from Kew.

---

* As far as I can make out, the history of our knowledge on tendrils is as follows :—We have seen that Palm and Von Mohl observed about the same time the singular phenomenon of the spontaneous revolving movement of twining-plants. Palm (S. 58), I presume, observed likewise the revolving movement of tendrils ; but I do not feel sure of this, for he says very little on the subject. Dutrochet fully described this movement of the tendril in the common Pea. Mohl first discovered that tendrils were sensitive to contact; but from some cause, probably from observing too old tendrils, he was not aware how sensitive they were, and thought that prolonged pressure was necessary to excite movement. Professor Asa Gray, in a paper already quoted, first noticed the extreme sensitiveness and rapidity of movements in the tendrils of certain Cucurbitaceous plants.

† This and the following drawings, from which the woodcuts have been engraved, were carefully made for me from living plants by my son Mr. George H. Darwin.

E

degree of spontaneous movement; for in one case they certainly described minute, irregular, vertical ellipses. The tendrils apparently curve themselves spontaneously to the same side with the petioles; but the movement was so slight that it may be passed over. From various causes, it was difficult to observe the movements of the petioles and tendrils in this and the two following species. The tendrils are so closely similar in all respects to those of the following species, that one description will suffice.

Bignonia unguis.—The young shoots revolve, but less regularly and less quickly than those of the last species. The stem twined imperfectly round a vertical stick, sometimes reversing its direction, exactly in the same manner as has been described in so many leaf-climbers; and this plant is in itself a leaf-climber, though possessing tendrils. Each leaf consists of a petiole bearing a pair of leaflets, and terminating in a tendril, which is exactly like that above figured, but a little larger. The whole tendril in a young plant was only about half an inch in length, and is very unlike most tendrils in shape. It curiously resembles the leg and foot of a small bird with the hind toe cut off. The straight leg or tarsus is longer than the three toes, which latter are of equal length, and, diverging, lie in the same plane; the toes terminate in sharp and hard claws, much curved downwards, exactly like the claws on a bird's foot. The whole tendril apparently represents three leaflets. The main petiole (but not the two sub-petioles of the lateral leaflets) is sensitive to contact with any object: even a small loop of thread after two days caused one to bend upwards. The whole tendrils, namely the tarsus and three toes, especially their under surfaces, are likewise sensitive to contact. Hence, when a shoot grows through branched twigs, its revolving movement soon brings the tendril into contact with some twig, and then all three toes bend (or sometimes one alone), and, after several hours, seize fast hold of the twig, exactly like a bird when perched. The tarsus, also, when it comes into contact with a twig, slowly bends, until the foot is carried quite round, and the toes pass on each side of the tarsus, or seize hold of it. If the main petiole bearing the leaflets comes into contact with a twig, it likewise bends round, until the tendril touches its own petiole or that of the opposite leaf, which is then seized. The petioles, and probably even the tendrils in a slight degree, move spontaneously; hence when a shoot attempted to twine round an upright stick, both petioles after a time came into contact with it, and the contact caused still further bending; so that ultimately both petioles clasped the stick in opposite directions, and the foot-like tendrils, seizing on each other or on their petioles, fastened the stem to the support with surprising security. Hence this species, differently from the last, uses its tendrils, by the intervention of the spontaneously moving and sensitive petioles, when the stem twines round a thin vertical stick. Both species use their tendrils in the same manner when passing through a thicket. This plant seems to me the most efficient climber which I have examined; and it probably could ascend a polished stem incessantly tossed by heavy storms. To show how important vigorous health is for the action of all the parts, I may mention that when I first examined a plant which was growing pretty well, though not vigorously, I concluded that the tendrils acted only like the hooks on a bramble, and that this was the most feeble and inefficient of all climbers!

Bignonia Tweedyana.—This species is closely allied to, and behaves in all respects like the last; perhaps it twines round a vertical stick rather better. On the same plant, one branch twined in one direction and another in an opposite direction. The internodes in one case made two circles, each in 2 h. 33 m. I was enabled in this species to observe, better than in the two preceding, the spontaneous movements of the petioles : one described three small vertical ellipses in the course of eleven hours, another moved laterally in an irregular spire. Some little time after the stem has twined round an upright stick, and is securely fastened to it by the clasping petioles and tendrils, it emits at the base of its leaves aërial roots, which curve partly round and adhere to the stick; so that this one species of *Bignonia* combines four different methods of climbing, generally characteristic of distinct plants, namely, twining, leaf-climbing, tendril-climbing, and root-climbing.

In the foregoing three species, when the foot-like tendril has caught any object, it continues to grow and to thicken, and ultimately it becomes wonderfully strong, in the same manner as we have seen with the petioles of leaf-climbers. If the tendril catches nothing, it first slowly bends downwards, and then its power of clasping is lost. Very soon afterwards it disarticulates itself from the petiole, like a leaf in autumn from the stem, and drops off. I have seen this process of disarticulation in no other tendrils, but when uncaught they soon wither away.

Bignonia venusta.—The tendrils are here considerably modified

E 2

in comparison with those of the previous species. The lower part, or tarsus, is four times as long as the three toes ; these are of equal length ; they do not lie in the same plane, but diverge equally on all sides ; their tips are bluntly hooked, so that the whole tendril makes an excellent grapnel. The tarsus is sensitive on all sides ; but the three toes are sensitive only on their outer surfaces, which correspond with the under surfaces of the toes in the tendrils of the previous species. The sensitiveness is not much developed ; for a slight rubbing with a twig did not cause the tarsus or toes to become slightly curved until an hour had elapsed ; subsequently they straightened themselves. Both tarsus and toes can seize well hold of sticks. When the stem is secured, the tendrils are seen spontaneously to sweep large ellipses : the two opposite tendrils move independently of each other. I have no doubt, from the analogy of the two following allied species, that the petioles move spontaneously ; but they are not irritable like those of *B. unguis* and *B. Tweedyana.* The young internodes also sweep fine large circles, one being completed in 2 h. 15 m., and a second in 2 h. 55 m. By these combined movements of the internodes, petioles, and grapnel-like tendrils, the latter are soon brought into contact with surrounding objects. When a shoot stands near an upright stick, it twines regularly and spirally round it ; as it ascends, it seizes the stick with only one of its tendrils, and, if the stick be thin, the right- and left-hand tendrils are alternately used. This alternation follows from the stem necessarily taking one twist round its own axis for each completed spire.

The tendrils a short time after catching any object contract spirally. Those which have caught nothing slowly bend downwards, but do not contract spirally. With many plants the tendrils after a time contract spirally, whether or not they have caught any object. But this whole subject of the spiral contraction of tendrils will be discussed after the several tendril-bearing plants have been described.

Bignonia littoralis.—The young internodes revolve in fine large ellipses. An internode bearing immature tendrils made two revolutions, each in 3 h. 50 m. ; but when grown older, with the tendrils mature, two ellipses were performed, each at the rate of 2 h. 44 m. But this species, unlike the preceding, is incapable of spirally twining round any object : this did not appear due to any want of flexibility in the internodes, or to the action of the tendrils, and certainly not to any want of the revolving power ; nor can I account for the circumstance. Nevertheless the plant readily ascends a thin upright stick by its two opposite tendrils, both seizing the stick some way above, and afterwards spirally contracting. If the tendrils seize nothing, they do not contract spirally. *Bignonia venusta* ascended a vertical stick by spirally twining and by seizing it alternately with its two tendrils like a sailor pulling himself up a rope hand over hand; our present species pulls itself straight up, like a sailor seizing with both hands together the rope above his head.

The tendrils are almost identical in structure with those of the last species. They continue growing for some time, even after clasping an object, and when fully grown, though borne by a young plant, were 9 inches in length. The three divergent toes are shorter relatively to the tarsus than in the former species; they are blunt at their tips and but slightly hooked; they are not quite equal in length, one being rather longer than the others. The outer surfaces of the three toes are highly sensitive; for when lightly rubbed with a twig, they became perceptibly curved in 4 m. and greatly curved in 7 m.; in 7 h. they became straight again and ready to react. The tarsus, for a space of one inch close to the toes, is sensitive, but in a rather less degree than the toes; for after a slight rubbing this part required about twice as long a time to bend. Even the middle part of the tarsus, if acted on soon after the tendril has arrived at maturity, is sensitive to prolonged contact. After the tendrils have grown old, the sensitiveness is confined to the toes, when they will only curl very slowly round a stick. The maturity of the tendril is shown by the divergence of the three toes, at which period their outer surfaces first become irritable. The irritability of the tendril has little power of spreading from one part to another: thus, when a stick was caught by the part immediately beneath the three toes, these often remained sticking out, and never clasped the stick.

The tendrils revolve spontaneously. The movement begins before the tendril is converted into a grapnel by the divergence of the toes, and before any part has become sensitive; so that the revolving movement is at this early period quite useless. The movement is at this time slow, two ellipses being completed conjointly in 24 h. 18 m. When the tendril was mature, an ellipse was performed in 6 h.; so that even at this period the movement is much slower than that of the internodes. Large ellipses were swept, both in vertical and horizontal planes, by the tendrils.

Not only the tendrils, but the petioles bearing them, revolve; these petioles, however, are not in the least sensitive. Thus the young internodes, the petioles, and the tendrils, all at the same time, go on revolving together, but at different rates. Moreover, the movements of the opposite petioles and tendrils are quite independent of each other. Hence, when the whole shoot is allowed freely to revolve, nothing can be more intricate than the course and rate followed by the extremity of each tendril. A wide hemisphere above the shoot is irregularly searched for some object to be grasped.

One other curious point remains to be mentioned. Some few days after the toes have closely clasped a stick, their blunt extremities become, though not invariably, developed into irregular disk-like balls, which have the singular power of adhering firmly to the wood. As similar cellular outgrowths will be fully described under *B. capreolata*, I will here say nothing more about them.

Bignonia æquinoctialis, var. *Chamberlaynii.*—The internodes, the elongated non-sensitive petioles, and the tendrils all have the power of revolving. The stem does not twine, but ascends a vertical stick in the same manner as the last species. The tendrils resemble those of the last species, but are shorter; the three toes are more unequal in length, two of them being about one-third shorter, and rather thinner than the third; but they vary in these respects. They terminate in small hard points; and what is important, they do not develope cellular adhesive disks. The reduced size of two of the toes, and their lessened sensitiveness, seem to indicate a tendency to their abortion; and the first-formed tendrils on one of my plants were sometimes quite simple. We are thus naturally led to the three following species with simple undivided tendrils.

Bignonia speciosa.—The young shoots revolve irregularly, making narrow ellipses, or spires or circles, at rates varying from 3 h. 30 m. to 4 h. 40 m.; but the plant shows no tendency to twine. Whilst very young and not requiring any support it does not produce tendrils. The tendrils of a rather young plant were five inches in length; they revolve spontaneously, as do the short and not sensitive petioles. The tendrils, when rubbed, slowly bend to the rubbed side, and subsequently straighten themselves; but they are not highly sensitive. There is something strange in their action : I repeatedly placed upright, thick and thin, rough and smooth sticks and posts, and string suspended vertically, near them ; but these objects were not well seized. The tendrils, after clasping an upright stick, repeatedly loosed it again ; often they would not seize it at all, or their extremities did not coil closely round it. I have observed hundreds of tendrils in Cucurbitaceous, Passifloraceous, and Leguminous plants, and never saw one behave in this manner. When, however, my plant had grown to a height of eight or nine feet, the tendrils acted much better ; and one or both regularly seized an adjoining, thin, upright stick, not high up as with the three previous species, but in a nearly horizontal plane ; thus the non-twining stem was enabled to ascend the stick.

The simple undivided tendril ends in an almost straight, sharp, uncoloured point. The whole terminal part exhibits one odd habit, which in an animal would be called an instinct ; for it continually searches for any little dark hole into which to insert itself. I had two young plants ; and, after having observed this habit, I placed near them posts, which either had been bored by beetles, or which had become fissured in drying. The tendrils, by their own movement and by that of the internodes, slowly travelled over the surface of the wood, and when the apex came to a hole or fissure it inserted itself ; for this purpose the terminal part, half or quarter of an inch in length, often bent itself at right angles to the basal part. I have watched this process between twenty and thirty times. The same tendril would frequently withdraw from one hole and insert its point into a second one. I have seen a tendril keep its point in one instance for 20 h. and in another instance for 36 h. in a minute hole, and then withdraw it.

Whilst the point of a tendril is thus temporarily inserted, the opposite tendril goes on revolving. The whole length of a tendril often fits itself closely to the surface of the wood with which it is in contact ; and I have seen a tendril bend at right angles and place itself in a wide and deep fissure, with the apex again abruptly bent and inserted into a minute lateral hole. After a tendril has clasped a stick, it contracts spirally ;. if it catches nothing, it does not contract. When it has adapted itself to the inequalities of a thick post, though it has clasped nothing, or when it has inserted its apex into some little fissure, the stimulus suffices to induce spiral contraction ; and this contraction always draws the tendril away from the post. So that in every case the above-described nicely adapted movements were absolutely useless, excepting once when the tip became jammed in a narrow

fissure. I fully expected, from the analogy of *B. capreolata* and
B. littoralis, that the tip would have developed itself into an
adhesive disk; but I could never detect even a trace of this
process. Improbable as the view may be, I am led to suspect that
this habit in the tendril of inserting its tip into dark holes and
crevices has been inherited by the plant after having lost the power
of forming adhesive disks.

Bignonia picta.—This species closely resembles the last in the
structure and movements of its tendrils: I casually examined a
fine growing plant of the allied *B. Lindleyi*, and this apparently
behaves in all respects in the same manner.

Bignonia capreolata.—We now come to a species having tendrils of a different type : but first for the internodes. A young
shoot made three large revolutions, following the sun, at an
average rate of 2 h. 23 m. The stem is thin and flexible, and I have
seen one make four regular spiral turns round a thin upright stick,
ascending, of course, from right to left, and therefore in a reversed
direction compared with the first-described species ; but afterwards, from the interference of the tendrils, it ascended either
straight up the stick or in an irregular spire. These tendrils are
highly remarkable. In a young plant they were about $2\frac{1}{2}$ inches
in length, and much branched, the five chief branches apparently
representing two pairs of leaflets and a terminal one ; each branch
is bifid or more commonly trifid toward its extremity, with all the
points blunt but distinctly hooked. A tendril when lightly rubbed bends to that side, and subsequently becomes straight again ;
but a loop of thread weighing $\frac{1}{4}$th of a grain produced no effect.
The terminal branches of a tendril twice became in 10 m. slightly
curved when touching a stick ; and in 30 m. the tips curled quite
round the stick: the basal part is less sensitive. The tendrils
revolve in an apparently capricious manner, sometimes not at all,
or very slightly, but at other times they describe large regular
ellipses. I could detect no spontaneous movement in the petioles.

At the same time that the tendrils are revolving more or less
regularly, another remarkable movement first begins ; the tendrils
slowly begin to bend from the light towards the darkest side of the
house. I repeatedly changed the position of my plants, and the
successively formed tendrils always ended by pointing, some little
time after the revolving movement had quite ceased, to the darkest
side. But when I placed a thick post near a tendril, and between
it and the light, the tendril pointed in that direction. In two instances a pair of leaves stood so that one tendril was directed to-

wards the light and the other to the darkest side of the house ; the latter did not move, but the opposite one bent itself first upwards and then right over its fellow, so that the two became parallel, one above the other, both pointing to the dark : I then turned the plant half round ; and the tendril which had turned over recovered its original position, and the opposite one, which had not moved before, now turned right over to the dark side. Lastly, on another plant, three pairs of tendrils were produced by three shoots at the same time, and all happened to be differently directed : I placed the pot in a box open only on one side, and obliquely facing the light ; in two days all six tendrils pointed with unerring truth to the darkest corner of the box, though to do this each had to bend in a different manner. Six tattered flags could not have pointed more truly from the wind than did these branched tendrils from the stream of light which entered the box. I left these tendrils undisturbed for above 24 h., and then turned the pot half round ; but they had now lost the power of movement, so that they could not any longer avoid the light.

When a tendril has not succeeded, either through its own re- volving movement or that of the shoot, or by turning towards any object which intercepts the light, in clasping a support, it bends vertically downwards and then towards its own stem, which it seizes together with the supporting stick, if there be one. A little aid is thus given in keeping the stem secure. If the tendril seizes nothing, it does not contract spirally, but soon withers away and drops off. If it does seize an object, all its branches contract spirally.

I have stated that, after a tendril has come into contact with a stick, in about half an hour it bends round it ; but I repeatedly observed, as with *B. speciosa* and its allies, that it again loosed the stick : sometimes it seized and loosed the same stick three or four times. Knowing that the tendrils avoided the light, I gave them a glass tube blackened within, and a well-blackened zinc plate : the branches curled round the tube and abruptly bent them- selves round the edges of the zinc plate ; but they soon recoiled, with what I can only call disgust, from these objects, and straight- ened themselves. I then placed close to a pair of tendrils a post with extremely rugged bark ; twice the tendrils touched it for an hour or two, and twice they withdrew ; at last one of the hooked extremities curled round and firmly seized an excessively minute projecting point of bark, and then the other branches spread themselves out, following with accuracy every inequality of the surface. I then placed a post without bark, but much fissured, and the points of the tendrils crawled into all the crevices in a beautiful manner. To my surprise, I observed that the tips of immature tendrils, with the branches not yet fully separated, likewise crawled, just like roots, into the minutest crevices. In two or three days after the tips had thus crawled into the crevices, or after their hooked ends had seized some minute point, the final process, now to be described, commenced.

This process I discovered by having accidentally left a piece of wool near a tendril. I then bound a quantity of flax, moss, and wool (the wool must not be dyed, for these tendrils are excessively sensitive to some poisons) loosely round sticks, and placed them near tendrils. The hooked points soon caught the fibres, even loosely floating fibres, and now there was no recoiling; on the contrary, the excitement from the fibres caused the hooks to penetrate the fibrous matter and to curl inwards, so that each hook firmly caught one or two fibres, or a small bundle of them. The tips and the inner surfaces of the hooks now began to swell, and in two or three days could be seen to be visibly enlarged. After a few more days the hooks were converted into whitish, irregular balls, rather above the $\frac{1}{20}$th of an inch in diameter, and formed of coarse cellular tissue, which sometimes wholly enveloped and concealed the hooks themselves. The surfaces of these balls secrete some viscid resinous matter, to which the fibres of the flax, &c. adhere. When a fibre has become fastened to the surface, the cellular tissue does not grow directly beneath it, but continues to grow closely on each side; so that when several adjoining fibres, though excessively thin, were caught, so many crests of cellular matter, each not as thick as a human hair, grew up between them, and these, arching over on both sides, grew firmly together. As the whole surface of the ball continues to grow, fresh fibres adhere and are enveloped; so that I have seen a little ball with between fifty and sixty fibres of flax crossing at various angles, all imbedded more or less deeply. Every gradation in the process could be seen—some fibres merely sticking to the surface, others lying in more or less deep furrows, or deeply imbedded, or passing through the very centre of the cellular ball. The imbedded fibres are so closely clasped that they cannot be withdrawn. The cellular outgrowth has such a tendency to unite, that two balls produced from two branches sometimes grow into a single one.

On one occasion, when a tendril had curled round a small stick, half an inch in diameter, an adhesive disk was formed; but generally the tendrils can do nothing with smooth sticks or posts. If, however, the tip of any one branch can curl round the minutest projecting point, the other branches will form disks, especially if they can find crevices to crawl into. The tendril quite fails to attach itself to a brick wall.

I infer that the disks or balls secrete some resinous adhesive matter, from the adherence of the fibres to them, but more especially from such fibres becoming loose after immersion in sulphuric ether, which likewise removes small, brown, glistening points that can generally be seen on the surface of the older disks. If the hooked extremities of the tendrils touch nothing, the cellular outgrowth, as far as I have seen, never commences; but temporary contact during a moderate time causes small disks to be formed. I have seen eight disks developed on one tendril. After the development of the disks, the tendrils, which now become spirally contracted, likewise become woody and very strong. A tendril in this state supported nearly seven ounces, and would apparently have supported a considerably greater weight had not the fibres of flax to which the disks were attached yielded.

From the facts above given, I infer that though the tendrils of this *Bignonia* can occasionally adhere to smooth cylindrical sticks and often to rugged bark, yet that they are specially adapted to climb trees clothed with lichens, mosses, or with *Polypodium incanum*, which I hear from Professor Asa Gray is the case with the forest-trees where this *Bignonia* grows. Finally, it is a highly remarkable fact that a leaf should become metamorphosed into a branched organ which turns from the light, and which can by its extremities either crawl like roots into crevices, or seize hold of minute projecting points, these extremities subsequently forming cellular masses which envelope by their growth the finest fibres and secrete an adhesive cement.

Eccremocarpus scaber (*Bignoniaceæ*).—Plants in the greenhouse, though growing pretty well, showed no spontaneous movements in their shoots or tendrils; but, removed to the hot-house, the young internodes revolved at rates varying from 3 h. 15 m. to 1 h. 13 m.: at this latter unusually quick rate one large circle was swept; but generally the circles or ellipses were small, and sometimes the course pursued was extremely irregular. An internode which had made several revolutions would sometimes stand

quite still for 12 h. or 18 h., and then recommence revolving; such
strongly marked interruptions in the movements I have observed
in no other plant.

The leaves bear four leaflets, themselves subdivided, and terminate in a much-branched tendril. The main petiole of the leaf,
whilst young, moves spontaneously by curving itself, and follows
nearly the same irregular course, and at about the same rate, with
the internodes. The movement to and from the stem is naturally
the most conspicuous, and I have seen the chord of the curved
petiole forming an angle of 59° with the stem, and an hour afterwards an angle of 106°. The two opposite petioles do not move
together, and one is sometimes raised so much as to stand close to
the stem whilst the other is not far from horizontal. The basal part
of the petiole moves less than the distal part. The tendrils, besides being carried by the moving petioles and internodes, themselves move spontaneously, and the opposite tendrils occasionally
move in opposite directions. By these several movements of the
young internodes, of the petioles, and of the tendrils, all acting
together, a wider space is swept for a support.

In young plants, the tendrils are about three inches in length:
they bear two lateral and two terminal branches; and each branch
bifurcates twice, with the tips forming blunt double hooks, having
both points directed to the same side. All the branches are sensitive on all sides; and after being lightly rubbed, or after coming
into contact with a stick, they bend in about 10 m. One that became, after a light rub, curved in 10 m., continued bending for
between 3 h. and 4 h., but subsequently in 8 h. or 9 h. became
straight again. Tendrils, which have caught nothing, ultimately
contract into an irregular spire, as they do also, only much more
quickly, after clasping a support. In both cases the petiole bearing the leaflets, which at first is straight and inclined a little
upwards, moves downwards and abruptly bends itself in the middle
into a right angle; but this is more plainly seen in *E. miniatus*
than in *E. scaber*. The action of the tendrils in the *Eccremocarpus* is in some respects analogous to that of the tendrils of
Bignonia capreolata; but the whole tendril does not move from
the light, nor do the hooked tips become enlarged into cellular
disks. After the tendrils have come into contact with moderately
thick cylindrical sticks or with rugged bark, the several branches
may be observed slowly to lift themselves up, change their position, and again come into contact with them. The object of these

movements is that the double hooks at the extremities of the branches, which naturally face in all directions, may be brought into contact with the wood. I have watched a tendril, which had bent itself at right angles abruptly round the sharp corner of a post, neatly bring every single hook into contact with both surfaces. The appearance suggested the belief, that though the whole tendril is not sensitive to light, yet that the tips are so, and that they turn and twist themselves towards any opaque surface. Ultimately the branches arrange and fit themselves very neatly to all the irregularities of the most rugged bark, so that they resemble in their irregular course a river with its branches, as engraved on a map. But when a tendril has thus arranged itself round a rather thick smooth stick, the subsequent spiral contraction generally spoils the neat arrangement, and draws the tendril from its support. So it is, but not in quite so marked a manner, when a tendril has spread itself over the rugged bark of a thick trunk; for in this case the spiral contraction of the opposite branches sometimes draws the opposed hooks firmly to their supports. Hence we may conclude that these tendrils are not perfectly adapted to seize smooth moderately thick sticks or rugged bark. When a thin stick or twig is placed near a tendril, its terminal branches wind quite round it and seize their own lower branches or main stem; and the stick is thus firmly, but not neatly, grasped. The extremities of the branches, close to the little double hooks, have a strong tendency to curl inwards, and are excited to this movement by contact with the thinnest objects. This accounts for the tendrils apparently preferring such objects as excessively thin culms of a grass, or the long flexible bristles of a brush, or the thin rigid leaves of an Asparagus, all which objects they seized in an admirable manner; for the tips of each sub-branch seized one, two, or three of the bristles, for instance, and then the spiral contraction of the several branches brought all these little parcels close together, so that thirty or forty bristles were drawn into a single bundle, and afforded an excellent support.

POLEMONIACEÆ.—*Cobæa scandens*.—This is an admirably constructed climber. The terminal portion of the petiole, which forms the tendril, was in one very fine specimen eleven inches in length, with the basal part bearing two pairs of leaflets, only two and a half inches in length. The tendril of the *Cobæa* revolves more rapidly and vigorously than in any other plant observed by me, with the exception of one *Passiflora*. It made three fine large, nearly circular sweeps, against the sun, each in 1 h. 15 m., and two others in 1 h. 20 m. and 1 h. 23 m. Sometimes it travels in a much inclined position, and sometimes nearly upright. The lower part moves but little, and the basal portion or petiole, which bears the leaflets, not at all; nor do the internodes revolve; so that here we have the tendril alone moving. With most of the species of *Bignonia* and with *Eccremocarpus*, the internodes, tendrils, and petioles all revolve. The long, straight, tapering main stem of the tendril of the *Cobæa* bears alternate branches; and each branch is several times divided, with the finer branches as thin as very thin bristles, extremely flexible, so that they are blown about by a breath of air, yet strong and highly elastic. The extremity of each branch is a little flattened, and terminates in a minute double (but sometimes single) hook, formed of hard, transparent, woody substance, and as sharp as the finest needle. On the eleven-inch tendril I counted ninety-four of these beautifully constructed little hooks. They readily catch soft wood, or gloves, or the skin of the hands. Excepting these hardened hooks, and excepting the basal part of the central stem of the tendril, every part of every branch is highly sensitive on all sides to a slight touch, and bends in a few minutes towards the touched side. By lightly rubbing several branches on different and opposite sides, the whole tendril rapidly assumes an extraordinarily crooked shape: these movements from contact do not interfere with the ordinary revolving movement. The branches, after becoming greatly curved from being touched, straighten themselves at a quicker rate than in almost any other tendril seen by me, namely, in between half an hour and an hour. After the tendril has caught any object, the spiral contraction also begins after an unusually short interval of time, namely, in about twelve hours.

Before the tendril is mature, the terminal branches cohere and the hooks are curled closely inwards: at this period no part is sensitive to a touch; but as soon as all the branches have diverged and the hooks stand out, full sensitiveness is acquired. It is a singular circumstance that the immature tendril, before becoming sensitive, begins to revolve at its full velocity: this movement must be useless, as the tendril in this state can catch nothing: it is a rare instance of a want, though only for a short time, of perfect co-adaptation in the structure and functions of a climbing-plant. The petiole with the tendril perfectly matured, but with the leaflets still quite small, stands at this period vertically upwards, the young growing shoot or axis being thrown to one side. The tendril thus standing vertically up sweeps a circle right above the stem, and is well adapted to catch some object above, and to favour the ascent of the plant. The whole leaf, with its tendril, after a short time, bends downwards to one side, allowing the next succeeding leaf to become vertical, and ultimately it assumes a horizontal position; but, before this has occurred, the tendril, supposing it to have caught nothing, has lost its powers of movement and has spirally contracted into an entangled mass. In accordance with the rapidity of all the movements, their duration is short : in a plant growing vigorously from being placed in a hot-house, a tendril only revolved for about 36 hours, counting from the period when it became sensitive; but during this period it probably made at least 27 revolutions.

When the branches of a revolving tendril strike against a stick, they quickly bend round and clasp it; but the little hooks play an important part, especially if only the extremity of the tendril be caught, in preventing its being dragged by the rapid revolving movement away too quickly for its irritability to act. As soon as a tendril has bent round a smooth stick or a thick rugged post, or has come into contact with planed wood (for it can at least temporarily adhere even to so smooth a surface as this), the same peculiar movements begin in the branchlets as have been described in those of the *Bignonia capreolata* and the *Eccremocarpus*, namely, the branchlets lift themselves up and down; those, however, which have their hooks already directed downwards remain in this position and secure the tendril, whilst the others twist about till they arrange themselves in conformity with every irregularity of the surface, and bring their hooks, originally facing in various directions, into contact with the wood. The use of the hooks was shown by giving the tendrils tubes and slips of glass to catch; for these, though temporarily seized, were afterwards invariably lost, either during the arrangement of the branches or when the spiral contraction ensued.

The perfect manner in which the branches arrange themselves, creeping like rootlets over all the inequalities and into any deep crevice, is quite a pretty sight; for it is perhaps more effectually done than by the tendrils of the former species, and is certainly more conspicuous, as the upper surfaces of the main stem and of every branch to the extreme hooks are angular and coloured green, whilst the lower surfaces are rounded and purple. I was led to infer, as in the former cases, that light guided these conforming movements of the branches of the tendrils. I made

many trials with black and white glass and cards to prove it, but
failed from various causes; yet these trials countenanced the
belief. The tendril may be looked at as a leaf split into filaments,
with the segments facing in all directions; hence, when the revolving movement is arrested, so that the light shines on them
steadily in one direction, there is nothing surprising in their
upper surfaces turning towards the light: now this may aid, but
will not account for, the whole movement; for the segments would
in this case move towards the light as well as turn round to it,
whereas in truth the segments or branches of the tendrils not only
turn their upper surfaces to the light, and their lower surfaces
which bear the hooks to any closely adjoining opaque object (that
is, to the dark), but they actually curve or bend from the light
towards the dark.

When the *Cobæa* grows in the open air, the wind must aid the
extremely flexible tendrils in seizing a support, for I found a mere
breath sufficed to cause the extreme branches of a tendril to catch
by their hooks twigs which they could not have reached by the
revolving movement. It might have been thought that a tendril
thus hooked only by its extremity could not have fairly grasped its
support. But several times I watched cases like the following,
one of which alone I will describe: a tendril caught a thin stick
by the hooks of one of its two extreme branches; though thus
held by the tip, it continued to try to revolve, bowing itself out to
all sides, and thus moving its branches; the other extreme branch
soon caught the stick; the first branch then loosed itself, and
then, arranging itself afresh, again caught hold. After a time, from
the continued movement of the tendril, a third branch became
caught by a single extreme hook; no other branches, as things then
remained, could possibly have touched the stick; but before long
the main stem, towards its extremity, began just perceptibly to
contract into an open spire, and thus to shorten itself (dragging
the whole shoot towards the stick), and as it continued to try to
revolve, a fourth branch was brought into contact. As the spiral
contraction travelled down the main stem and down the branches
of the tendril, all the lower branches, one after another, were
brought into contact with the stick, and were wound round it and
round their own branches until the whole was tied together in an
inextricable knot round the stick. The branches of a tendril,
though at first so flexible, after having clasped a support for
a time, become rigid and even stronger than they were at first.
Thus the plant is secured to its support in a perfect manner.

LEGUMINOSÆ.—*Pisum sativum*.—The common Pea was the subject of a valuable memoir by Dutrochet[*], who discovered that both the internodes and tendrils revolved in ellipses. The ellipses are generally very narrow, but sometimes approach to circles: I several times observed that the longer axis slowly changed its direction, which is of importance, as the tendril thus sweeps a much wider circuit. Owing to this change of direction, and likewise to the movement of the stem towards the light, the successive irregular ellipses generally form an irregular spire. I have thought it worth while to annex a tracing of the course pursued by the upper internode (the movement of the tendril being neglected) of a young plant from 8.40 A.M. to 9.15 P.M. The course was traced on a hemispherical glass placed over the plant, and the dots with figures give the hours of observation; each dot was joined by a straight line: no doubt these lines, if the course had been observed at shorter intervals, would have been all curvilinear. The extremity of the petiole, where the young tendril arises, was 2 inches from the glass, so that if a pencil 2 inches long had been in imagination affixed to the petiole, it would have traced the annexed figure on the under side of the glass; but it must be remembered that the figure is here reduced one-half. Neglecting the first great sweep towards the light or window, the end of the petiole swept a space 4 inches

Fig. 6.

Diagram showing the movement of the upper internodes of the common Pea, traced on a hemispherical glass and transferred to paper; reduced one-half in size. (Aug. 1st.)

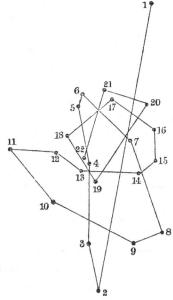

Side of room with window.

h. m.	h. m.	h. m.
1. 8 46 A.M.	9. 1 55 P.M.	16. 5 25 P.M.
2. 10 0 „	10. 2 25 „	17. 5 50 „
3. 11 0 „	11. 3 0 „	18. 6 25 „
4. 11 37 „	12. 3 30 „	19. 7 0 „
5. 12 7 P.M.	13. 3 48 „	20. 7 45 „
6. 12 30 „	14. 4 40 „	21. 8 30 „
7. 1 0 „	15. 5 5 „	22. 9 15 „
8. 1 30 „		

across in one direction, and 3 inches in another. As a full-grown tendril is considerably above 2 inches in length, and as the

* Comptes Rendus, tom. xvii. 1843, p. 989.

tendril itself bends and revolves in harmony with the internode, a considerably wider space than that here specified (and represented one-half reduced) is swept. Dutrochet observed an ellipse completed in 1h. 20m.; I saw one completed in 1h. 30m. The direction followed is variable, either with or against the sun.

Dutrochet asserts that the petiole of the leaf spontaneously moves, as well as the young internodes and tendrils; but he does not say that he secured the internodes; when this was done, I never detected any movement in the petiole, except to and from the light.

The tendrils, on the other hand, when the internodes and petioles were secured, described irregular spires or regular ellipses, exactly like those made by the internodes. A young tendril, only 1⅛ inch in length, revolved. Dutrochet has shown that when a plant is placed in a room, so that the light enters laterally, the internodes travel much quicker to the light than from it : on the other hand, he asserts that the tendril itself moves from the light towards the dark side of the room. With due deference to this great observer, I think he was mistaken, owing to his not having secured the internodes. I took a young plant with highly sensitive tendrils, and tied the petiole so that the tendril alone could move; it completed a perfect ellipse in 1h. 30m.; and I then turned the plant half round, so that the opposite side faced the light, but this made no change in the direction of the succeeding ellipse. The next day I watched a plant similarly secured until the tendril (which was highly sensitive) made an ellipse in a line exactly to and from the light ; the movement was so great that the tendril bent itself down at the two ends of its elliptical course into a line a little beneath the horizon, thus travelling more than 180 degrees ; but the curvature was fully as great towards the light as towards the dark side of the room. I believe Dutrochet was misled by not having secured the internodes, and by having observed a plant of which the internodes and tendrils, from inequality of age, no longer curved or moved in harmony together.

Dutrochet made no observations on the sensitiveness of the tendrils; these, whilst young and about an inch in length, with the leaflets on the petiole only partially expanded, are highly sensitive ; a single light touch with a twig on the inferior or concave surface near the tip caused them quickly to bend, as did occasionally a loop of thread weighing one-seventh of a grain. The upper or convex surface is barely or not at all sensitive. After bending from a touch the tendril straightened itself in about two hours, and was ready to act again. As soon as the tendrils begin to grow old their extremities become hooked, and they then appear, with their two or three pairs of branches, an admirable grappling instrument; but this is not really the case, for at this period the tips have generally quite lost their sensitiveness; when hooked on to twigs some were not at all affected, and others required from 18 h. to 24 h. to clasp the twigs. Ultimately the lateral branches of the tendril, but not the middle or main stem, contract spirally.

Lathyrus aphaca.—As the tendril here replaces the whole leaf (except occasionally in very young plants), the leaf itself being replaced in function by the large stipules, it might have been expected that the tendrils would have been highly organized; this, however, is not so. They are moderately long, thin, and unbranched, with their tips slightly curved : they are sensitive whilst young on all sides, but chiefly on the concave side of the extremity. They have no spontaneous revolving power, but are at first inclined upwards at an angle of about 45°, then move into a horizontal position, and ultimately bend downwards. The young internodes, on the other hand, revolve in ellipses, and carry with them the tendrils : two ellipses were completed, each in nearly 5 h.; the longer axes of these two, and of some subsequently formed ellipses, were directed at about an angle of 45° from the line of the axis of the previous ellipse.

Lathyrus grandiflorus.—The plants observed were young, and not growing vigorously, yet sufficiently so, I think, for my observations to be trusted. Here we have the rare case of neither internodes nor tendrils having any spontaneous revolving power. The tendrils in vigorous plants are above 4 inches in length, and are often twice divided into three branches; the tips are curved and are sensitive on the concave side ; the lower part of the central stem is hardly at all sensitive. Hence this plant climbs simply by its tendrils being brought, through the growth of the stem, or the more efficient aid of the wind, into contact with surrounding objects, which are then effectually clasped. I may add that the tendrils, or the internodes, or both, of *Vicia sativa* spontaneously revolve.

CompositÆ.—*Mutisia clematis.*—The enormous family of Compositæ is well known to include very few climbing plants. We have seen in the Table in the first Part that *Mikania* is a regular twiner, and *Mutisia* is the only genus, as far as I can learn, which bears tendrils : it is therefore interesting to discover that these tendrils, though rather less metamorphosed from their primordial

F 2

foliar nature than most other tendrils, yet display all the ordinary characteristic movements, both those that are spontaneous and those excited by contact.

The long leaf bears seven or eight alternate leaflets, and terminates in a tendril which, in a plant of considerable size, was 5 inches in length. It consists generally of three branches, which evidently represent in a much elongated condition the petioles and midribs of three leaflets; for the branches of the tendril are exactly like the petioles and midribs of the leaflets, being square on the upper surface, furrowed, and edged with green. Moreover, in the plant whilst quite young, the green edging to the branches of the tendrils sometimes expands into narrow laminæ or blades. Each branch is curved a little downwards, and is slightly hooked at its extremity.

An upper young internode revolved, judging from three revolutions, at an average rate of 1h. 38m.; it swept ellipses with the longer axes directed at right angles to each other; the plant, apparently, cannot twine. The petiole which bears the tendril, and the tendril itself, are both in constant movement. But the movement is slower and much less regularly elliptical than that of the internodes; it is, apparently, much affected by the light, for the whole leaf usually sank during the night and rose during the day, moving in a crooked course to the west. The tips of the tendrils are highly sensitive on their lower surfaces: one just touched with a twig became perceptibly curved in 3 m., and another became so in 5 m.; the upper surface is not at all sensitive; the sides are moderately sensitive, so that two branches rubbed on their adjoining sides converged and crossed each other. The petiole of the leaf and the lower part of the tendril, halfway between the upper leaflet and the lowest tendril-branch, are not sensitive. A tendril after curling from a touch became straight again in about 6 h., and was ready to react; but one that had been so roughly rubbed as to have coiled into a helix was not perfectly straight after 13 h. The tendrils retain their sensibility to an unusual age; for one borne by a leaf, with five or six fully developed leaves above it, was still active. If a tendril catches nothing, the tips of its branches, after a considerable interval of time, spontaneously curl a little inwards; but if the tendril has clasped some object, the whole length contracts spirally.

SMILACEÆ.—*Smilax aspera*, var. *maculata*.—Aug. St.-Hilaire[*] considers the tendrils which rise in pairs from the petiole as

* Leçons de Botanique, &c., 1841, p. 170.

modified lateral leaflets ; but Mohl (S. 41) ranks them as modified stipules. These tendrils are from 1½ to 1¾ inch in length, are thin, and have slightly curved, pointed extremities. They diverge a little from each other, but stand at first nearly upright. When lightly rubbed on either side, they slowly bend to that side, and subsequently become straight again. The back or convex side of a tendril placed in contact with a stick became just perceptibly curved in 1 h. 20 m., but did not completely surround the stick till 48 h. had elapsed ; the concave side of another tendril became considerably curved in 2 h., and fairly clasped the stick in 5 h. As the tendrils grow old, they diverge more from each other and slowly bend towards the stem and downwards, so that they project on the opposite side of the stem to that on which they arise ; they still retain their sensitiveness, and can clasp a support placed behind the stem. Owing to this movement, the plant can ascend a thin upright stick, clasping it with the tendrils which arise from the leaves placed alternately on opposite sides of the stem. Ultimately the two tendrils belonging to the

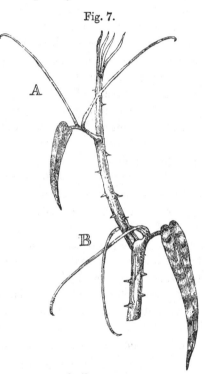

Fig. 7.

Smilax aspera.

same petiole, if they do not come into contact with any object, cross each other (as at B in fig. 7) behind the stem and loosely clasp it. This movement of the tendrils towards and round the stem is, to a certain extent, guided by the action of the light ; for when the plant stood so that one of the two tendrils in thus slowly moving had to travel towards the light, and the other from the light, the latter always travelled, as I repeatedly observed, more quickly than its fellow. The tendrils do not contract spirally in any case. Their chance of finding a support depends on the growth of the plant, on the wind, and on their own slow backward and downward movement, which is guided, to a certain extent, by the movement from the light or towards any dark object; for neither the internodes nor the tendrils have any proper revolving movement. From this latter circumstance, from the slow movements of the tendrils after contact (though their sensitiveness is retained for an unusual length of time), from their simple structure and shortness, this plant shows less perfection in its means of climbing than any other tendril-bearing plant observed by me. Whilst young and only a few inches in height, it does not produce any tendrils; and considering that it grows to only about 8 feet high, that the stem is zigzag, and is furnished, as well as the petioles, with spines, it is surprising that it should be provided with tendrils, comparatively inefficient though they be. The plant might have been left, one would have thought, to climb by the aid of its spines alone, like our brambles. But, then, it belongs to a genus some of the species of which are furnished with much longer tendrils; and we may believe that *S. aspera* is endowed with these organs solely from being descended from progenitors more highly organized in this respect.

FUMARIACEÆ. — *Corydalis claviculata.* — According to Mohl (S. 43), both the leaves and the extremities of the branches are converted into tendrils. In the specimens examined by me all the tendrils were certainly foliar, and it is hardly credible that the same plant should produce tendrils of such widely different homological natures. Nevertheless, from this statement by Mohl, I have ranked this *Corydalis* amongst tendril-bearers; if classed exclusively by its foliar tendrils, it would be doubtful whether it ought not to have been placed amongst leaf-climbers, with its allies, *Fumaria* and *Adlumia*. A large majority of its so-called tendrils still bear leaflets, though excessively reduced in size; some few of them may be properly designated as tendrils, for they are completely destitute of laminæ or blades. Consequently we here behold a plant in an actual state of transition from a leaf-climber to a tendril-bearer. Whilst the plant is young, only the outer leaves, but when full-grown all the leaves, have their extremities more or less perfectly converted into tendrils. I have examined specimens from one locality alone, viz. Hampshire; and it is not improbable that plants growing under different conditions might have their leaves a little more or less changed into true tendrils.

Whilst the plant is quite young, the first-formed leaves are not modified in any way, but those next formed have their terminal leaflets reduced in size, and soon all the leaves assume the structure represented in the following diagram. This leaf bore nine leaflets; the lower ones are much subdivided. The terminal portion of the petiole, about 1½ inch in length (above the leaflet (*f*)), is thinner and more elongated than the lower part, and may

Fig. 8.

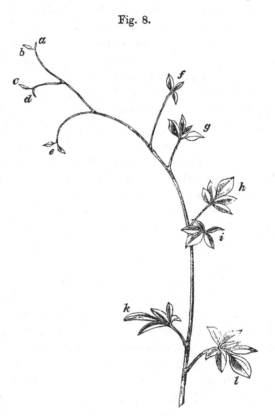

Corydalis claviculata.
Leaf-tendril, of natural size.

be considered as the tendril. The leaflets borne by this part are greatly reduced in size, being, on an average, about the tenth of an inch in length and very narrow; one small leaflet measured one-twelfth of an inch in length and one-seventy-fifth in breadth, so that it was almost microscopically minute. All the reduced leaflets have branching nerves, and terminate in little spines like the fully de-veloped leaflets. Every gradation can be traced, until we come to branchlets (as *a* and *d* in the figure) which show no vestige of a lamina or blade. Occasionally all the terminal branchlets of the petiole are in this latter condition, and we then have a true tendril.

The several terminal branches of the petiole bearing the much-reduced leaflets (*a, b, c, d*) are highly sensitive, for a loop of thread weighing only the one-sixteenth of a grain caused them, in under 4h., to become greatly curved : when the loop was removed, the petioles straightened themselves in about the same time. The petiole (*e*) was rather less sensitive; and in another specimen, in which the corresponding petiole bore rather larger leaflets, a loop of thread weighing one-eighth of a grain did not cause curvature until 18h. had elapsed. Loops of thread weighing one-fourth of a grain, left suspended on all the lower petioles (*f* to *l*) during several days, produced no effect. Yet the three petioles *f, g*, and *h* are not quite insensible, for when left in contact with a stick for a day or two they slowly curled round it. So that the sensibility of the petiole gradually diminishes from the tendril-like extremities to the base. The internodes are not at all sensitive, which makes Mohl's statement that they are sometimes converted into tendrils the more surprising, not to say improbable.

The whole leaf, whilst young and sensitive, stands almost vertically upwards, as we have seen is the case with many tendrils. It is in continual movement, and one that I observed swept large, though irregular, ellipses, sometimes narrow, sometimes broad, with their longer axes directed to different points of the compass, at an average rate of about 2h. for each revolution. The young internodes also, which bear the revolving leaves, likewise revolve irregularly in ellipses and spires; so that by these combined movements a considerable space is swept for a support. If the terminal and attenuated portion of the petiole fails in seizing any object, it ultimately bends downwards and inwards, and then soon loses all its irritability and power of movement. This bending down is of a very different nature from that which occurs with the extremities of the young leaves in many species of *Clematis*; for these, when thus bent or hooked, first acquire their full degree of sensitiveness.

Dicentra thalictrifolia.—In this allied plant the metamorphosis of the terminal leaflets has been complete, and they are converted into perfect tendrils. Whilst the plant was young, the tendrils appeared like modified branches, so that a distinguished botanist thought this was their nature ; but in a full-grown plant, there can be no doubt, as I am assured by Dr. Hooker, that the tendrils are modified leaves. The tendrils, when of full size, are above 5 inches in length; they bifurcate twice, thrice, or even four times; their extremities are hooked, but blunt. All the branches of the tendrils are sensitive on all sides, but the basal portion of the main stem is only slightly sensitive. The terminal branches lightly rubbed with a twig did not curve until from 30m. to 42m. had elapsed : they slowly became straight again in between 10h. and 20h. A loop of thread weighing one-eighth of a grain plainly caused the thinner branches to curve, as did occasionally a loop weighing one-sixteenth of a grain; but this latter slight weight, though left suspended, was not sufficient to cause a permanent flexure. The whole leaf with its tendril and the young upper internode together revolve vigorously and quickly, though irregularly, and sweep a wide space. The figure traced on a bellglass was either an irregular spire or a zigzag line. The nearest approach to an ellipse was an elongated figure of 8, with one end a little open; this was completed in 1h. 53m. During a period of 6h. 17m. another shoot made a complex figure, apparently representing three and a half ellipses. When the lower part of the petiole bearing the leaflets was securely fastened, the tendril itself described similar but much smaller figures.

This species climbs well. The tendrils after clasping a stick become thicker and more rigid; but the blunt hooks do not turn and adapt themselves to the supporting surface, as is the case in so perfect a manner with some of the Bignoniaceæ and the *Cobæa*. In young plants 2 or 3 feet in height, the tendrils, which are only half the length of those borne by the same plants when grown taller, do not contract spirally after clasping a support, but only become slightly flexuous. Full-sized tendrils, on the other hand, contract spirally, excepting the thick basal portion. Tendrils which have caught nothing simply bend downwards and inwards, like the extremities of the leaves of the *Corydalis claviculata*. But in all cases the petiole after a time becomes angularly and abruptly bent like that of the *Eccremocarpus*.

CUCURBITACEÆ.—The tendrils in this family have been ranked by several competent judges as modified leaves, stipules, and branches; or the same tendril as part leaf and part branch. De Candolle considers the tendrils in two of the tribes as different in their homological nature*. From facts recently adduced, Mr. Berkeley thinks that Payer's view is the most probable, namely, that the tendril is " a separate portion of the leaf itself "†.

* I am indebted to Prof. Oliver for information on this head. In the Bulletin de la Société Botanique de France, 1857, there are numerous discussions on the nature of the tendrils in this family.

† Gardeners' Chronicle, 1864, p. 721. From the affinity of the Cucurbitaceæ to the Passifloraceæ, it might be argued that the tendrils of the former are

Echinocystis lobata.—I made numerous observations on this plant (raised from seed sent me by Prof. Asa Gray), for here I first observed the spontaneous revolving movement of the internodes and of the tendrils ; and knowing nothing of the nature of these movements, was infinitely perplexed by the whole case, and by the false appearance of twisting of the axis. My observations may now be greatly condensed. I recorded thirty-five revolutions of the internodes and tendrils ; the slowest rate was 2 h., and the average, with no great fluctuations, was 1 h. 40 m. for each revolution. Sometimes I tied the internodes, so that the tendrils alone moved; at other times I cut off the tendrils whilst very young, so that the internodes revolved by themselves; but the rate was not thus affected. The course generally pursued was with the sun, but often in the opposite direction ; sometimes the movement during a short time would either stop or be reversed ; and this apparently resulted from the interference of the light, shortly after the plant was placed close to a window. In one instance, an old tendril, which had nearly ceased revolving, moved in one direction, whilst the young tendril above moved in the opposite direction. The two uppermost internodes alone revolve ; as the internodes grow old, the upper part alone moves. The summit of the upper internode made an ellipse or circle about 3 inches in diameter, whilst the tip of the tendril swept a circle 15 or 16 inches in diameter. During the revolving movement the internodes become successively curved to all points of the compass ; and often in one part of their course they were inclined, together with the tendril, at about 45° to the horizon, and in another part stood vertical. There was something in the appearance of the revolving internodes which continually gave the false impression that their movement was due to the weight of the long and spontaneously revolving tendril; but, on suddenly cutting off the tendril with a sharp scissors, the top of the shoot rose very little, and went on revolving : this false appearance is apparently due to the internodes and tendrils all curving and moving harmoniously together.

I repeatedly saw that the revolving tendril, though inclined during the greater part of its course at an angle of about 45° (in one case of only 37°) above the horizon, in one part of its course stiffened and straightened itself from tip to base, and became

---

modified flower-peduncles, as is certainly the case with the tendrils of Passion-flowers. Mr. R. Holland (Hardwicke's 'Science-Gossip,' 1865, p. 105) states that "a cucumber grew, a few years ago, in my own garden, where one of the short prickles upon the fruit had grown out into a long curled tendril."

nearly or quite vertical. This occurred both when the supporting internodes were free and when they were tied up; but was perhaps most conspicuous in the latter case, or when the whole shoot happened to stand in an inclined position. The tendril forms a very acute angle with the extremity of the shoot, which projects above the point where the tendril arises; and the stiffening always occurred as the tendril approached, and had to pass in its revolving course, the point of difficulty—that is, the projecting extremity of the shoot. Unless the tendril had the power of thus acting, it would strike against the extremity of the shoot, and be arrested by it. As soon as all three branches of the tendril have begun to stiffen themselves in this remarkable manner, as if by a process of turgescence, and to rise from an inclined into a vertical position, the revolving movement becomes more rapid; and as soon as the tendril has succeeded in passing the extremity of the shoot, its revolving motion, coinciding with that from gravity, often causes it to fall into its previously inclined position so quickly, that the end of the tendril could be distinctly seen travelling like the minute hand of a gigantic clock.

The tendrils are thin, from 7 to 9 inches in length, with a pair of short lateral branches rising not far from the base. The tip is slightly but permanently curved, so as to act to a limited extent as a hook. The concave side of the tip is highly sensitive to a touch, but not so the convex side, as was likewise observed by Mohl (S. 65) with other species of the family. I repeatedly proved this difference by lightly rubbing four or five times the convex side of one tendril, and only once or twice the concave side of another tendril, and the latter alone curled inwards: in a few hours afterwards, when those which had been rubbed on the concave side had recovered themselves, I reversed the process of rubbing, and always with a similar result. After touching the concave side, the tip becomes sensibly curved in one or two minutes; and subsequently, if the touch has been at all rough, it becomes coiled into a helix. But this helix will, after a time, uncoil itself, and be ready to act again. A loop of thin thread only one-sixteenth of a grain in weight caused a temporary flexure in a tendril. One of my plants had two shoots near each other, and the tendrils were repeatedly drawn across each other, but it is a singular fact that they did not once catch each other. It would appear as if the tendrils had become habituated to the contact of other tendrils, for the pressure thus caused would apparently be greater than that caused by a loop of soft thread weighing only the one-sixteenth of a grain. So it would appear that the tendrils are habituated to drops of water or to rain; for artificial rain made by violently flirting a wet brush produced not the least effect on them. I repeatedly rubbed rather roughly the lower part of a tendril, but never caused any curvature; yet this part is sensitive to prolonged pressure, for when it came into contact with a stick, it would slowly bend round it.

The revolving movement is not stopped by the extremity curling after having been touched. When one of the lateral branches of a tendril has firmly clasped any object, the middle branch continues to revolve. When a stem is bent down and secured, so that its tendril depends but is left free to move, its previous revolving movement is nearly or quite stopped; but it begins to rise in a vertical plane, and as soon as it has become horizontal the revolving movement recommences. I tried this four times; generally the tendril rose to a horizontal position in an hour or an hour and a half; but in one case, in which the tendril depended at an angle of 45° beneath the horizon, the movement took two hours; in another half-hour the tendril rose to 23° above the horizon and recommenced revolving. This upward vertical movement is independent of the action of light, for it took place twice in the dark, and another time with the light coming in on one side alone. The movement no doubt is guided by opposition to the force of gravity, as in the case of the ascent of the plumules of germinating seeds.

A tendril does not long retain its revolving power; as soon as this ceases, it bends downwards and contracts spirally. But after the revolving movement has ceased the tip still retains for a short time its sensitiveness to contact, but this can be of little service to the plant.

Though the tendril is highly flexible, and though the extremity travels, under favourable circumstances, at about the rate of an inch in two minutes and a quarter, yet its sensitiveness to contact is so great that it hardly ever fails to seize a thin stick placed in its path. The following case surprised me much: I placed a thin, smooth, cylindrical stick (and I repeated the experiment seven times) so far from a tendril, that its extremity could only curl half or three-quarters round the stick; but I always found in the course of a few hours afterwards that the tip had managed to curl twice or even thrice quite round the stick. I at first thought that this was due to rapid growth; but by coloured points and measurements I proved that there was no sensible increase of length by growth. When a stick, flat on one side, was similarly placed, the tip of the tendril could not curl beyond the flat surface, but coiled itself into a helix, which, turning to one side, lay flat on the little flat surface of wood. In one instance a portion of tendril three-quarters of an inch in length was thus dragged on to the flat surface by the coiling in of the helix. But the tendril thus acquires a very insecure hold, and generally slips off: in one case alone the helix subsequently uncoiled itself, and the tip then passed round and clasped the stick. The formation of a helix on the flat side of a stick apparently shows us that the continued striving of the tip to curl itself closely inwards gives the force which drags the tendril round a smooth cylindrical stick. In this latter case, whilst the tendril was slowly and quite insensibly crawling onwards, I several times observed through a lens that the whole surface was not in close contact with the stick; and I can understand the onward movement only by supposing that it is slightly vermicular, or that the tip alternately straightens itself a little and then again curls inwards, thus dragging itself onwards by an insensibly slow, alternate movement, which may be compared to that of a strong man suspended by the ends of his fingers to a horizontal pole, who works his fingers onwards until he can grasp the pole with the palm of his hand. However this may be, the fact is certain that a tendril which has caught a round stick by its extreme point can work itself onwards until it has passed twice or even thrice round the stick, and has permanently grasped it.

Hanburya Mexicana.—The young internodes and tendrils of this anomalous member of the family revolve in the same manner and at about the same rate with the *Echinocystis.* The stem does not twine, but can ascend an upright stick by the aid of its tendrils. The concave tip of the tendril is very sensitive; after rapidly coiling into a loop from a single touch, it straightened itself in 50m. The tendril, when in full action, stands vertically up, with the young projecting extremity of the shoot thrown a little on one side out of the way; but the tendril bears near its base, on the inner side, a short branch, which projects out at right angles, like a spur, with the terminal half bowed a little downwards. Hence, as the main vertical branch of the tendril revolves, the spur, from its position and rigidity, cannot pass over the extremity of the shoot in the same curious manner as do the three branches of the tendril of the *Echinocystis* by stiffening themselves at the proper point, but is pressed laterally against the young shoot in one part of the revolving course, and in another part is carried only a little way from it. Hence the sweep of the lower part of the tendril of the *Hanburya* is much restricted. Here a nice case of co-adaptation comes into play: in all the other tendrils observed by me the several branches become sensitive at the same period; had this been the case with the *Hanburya*, the rectangular spur-like branch being pressed, during the revolving movement, against the projecting end of the shoot, would infallibly have seized it in a highly injurious manner. But the main tendril, after revolving for a time in a vertical position, spontaneously bends downwards; and this, of course, raises the rectangular branch, which itself also curves upwards; so that by these combined movements the spur-like branch rises above the projecting end of the shoot, and can now move freely without touching it; then, and not until then, it first becomes sensitive.

The tips of both branches, when they come into contact with a stick, grasp it like any ordinary tendril. In a few days afterwards the inferior surface swells and becomes developed into a cellular layer, which adapts itself closely to the wood, and firmly adheres to it. This layer is analogous to the adhesive disks formed by the tips of the tendrils in some species of *Bignonia*, but in the *Hanburya* the layer is developed along the terminal portion of the tendril, sometimes for a length of $1\frac{3}{4}$ inch, but not at the extreme tip. The layer is white, whilst the tendril is green, and near the tip it could sometimes be seen to be thicker than the tendril itself; it generally spreads a little beyond the sides of the tendril, and its edge is fringed with free elongated cells, which have enlarged globular or retort-shaped heads. This cellular layer apparently secretes some resinous cement; for its adhesion to the wood was not lessened by immersion for 24 h. in alcohol or water, but was quite loosened by the action during the same period of ether and turpentine. After the tendril has once firmly coiled itself round a stick, it is difficult to imagine of what use the formation of the adhesive cellular layer can be. Owing to the spiral contraction, which ensues after a time, whether or not the tendril has clasped any object, it was never able to remain, excepting in one instance, in contact with a thick post or a nearly flat surface; if it could have become attached to such objects by means of the adhesive cellular layer, this layer would evidently have been of service to the plant. I hear from Dr. Hooker that several other Cucurbitaceous plants have adherent tendrils.

Of other Cucurbitaceæ, I observed in *Bryonia dioica*, *Cucurbita ovifera*, and *Cucumis sativa*, that the tendrils were sensitive and revolved; in the latter plant, Dutrochet* saw the movement of the tendril reversed; but whether the internodes as well as the tendrils revolve in these several species I did not observe. In *Anguria Warscewiczii*, however, the internodes, though thick and stiff, do revolve: in this plant the lower surface of the tendril, some time after clasping a stick, produces a coarsely cellular layer or cushion, fitting the wood, like that formed by the tendril of the *Hanburya*; but it was not in the least adhesive. In *Zanonia Indica*, which belongs to a different tribe of the family, both the forked tendrils and the internodes revolved, in periods between 2 h. 8 m. and 3 h. 35 m., moving against the sun.

VITACEÆ.—In this family and in the two following, namely, the Sapindaceæ and Passifloraceæ, the tendrils are modified flower-peduncles; so that they are axial in their nature. In this respect they differ from those of all the first described families, but perhaps not from those of the Cucurbitaceæ. The homological nature, however, of a tendril seems to make no difference in its action.

Vitis vinifera.—The tendril is thick and of great size; one from

Fig. 9.

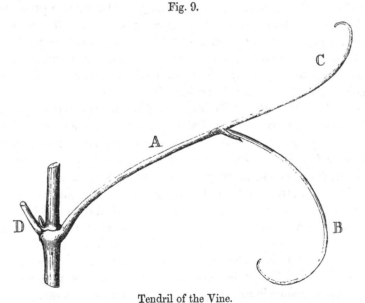

Tendril of the Vine.

A. Peduncle of tendril.   C. Shorter branch.
B. Longer branch, with a scale at its base.   D. Petiole of opposite leaf.

a vine not growing vigorously out of doors, measured 16 inches in length. It consists of a peduncle (A), bearing two branches

* Comptes Rendus, tom. xvii. p. 1005.

which diverge equally from it. One of the branches (B) has
a scale at its base, and is always, as far as I have seen, longer
than the other, and very often bifurcates. The several branches
when rubbed become curved, and subsequently straighten themselves. After a tendril has clasped any object by its extremity,
it contracts spirally; but this does not occur (Palm, S. 56) when
no object has been seized. The tendrils move spontaneously from
side to side; and on a very hot day one made two elliptical revolutions at an average rate of 2 h. 15 m. During these movements
a coloured line, painted along the convex surface, became first
lateral and then concave. The separate branches have independent movements; after a tendril has spontaneously revolved
for a time, it bends from the light towards the dark: I do not
give this latter statement on my own authority, but on that of
Mohl and Dutrochet; Mohl (S. 77) says that in a vine planted
against a wall the tendrils point towards it, and in a vineyard
generally more or less to the north.

The young internodes spontaneously revolve; but in hardly any
other plant have I seen so slight a movement. A shoot faced a
window, and I traced its course on the glass during two perfectly
calm and hot days; during ten hours on one day it described a
spire, representing two and a half ellipses. I likewise placed a
bell-glass over a young muscat grape in a hothouse, and it made
three or four extremely minute oval revolutions each day: the
shoot moved less than half an inch from side to side; and had it
not made at least three revolutions during the same day when the
sky was uniformly overcast, I should have attributed the motion
to the varying action of the light. The extremity of the shoot is
more or less bent downwards; but the extremity never reverses
its curvature, as so generally occurs with twining plants.

Various authors (Palm, S. 55; Mohl, S. 45; Lindley, &c.)
believe that the tendrils of the vine are modified flower-peduncles.
I here give a drawing (fig. 10) of the ordinary state of a flower-
peduncle in bud: it consists of the "common peduncle" (A); of
the "flower-tendril" (B), which is represented as having caught
a twig; and of the "sub-peduncle" (C) bearing the flower-buds.
The whole peduncle moves spontaneously, like a true tendril, but
in a less degree, and especially when the sub-peduncle (C) does
not bear many flower-buds. The common peduncle (A) has not
the power of clasping a support, nor has the corresponding part
in the true tendril. The flower-tendril (B) is always longer than
the sub-peduncle (C), and has a scale at its base; it sometimes

bifurcates, and therefore corresponds in every detail with the longer scale-bearing branch (B, fig. 9) of the true tendril. It is, however, inclined backwards from the sub-peduncle (C), or stands at right angles with it, and is thus adapted to aid in carrying the future bunch of grapes. The flower-tendril (B), when rubbed, curves and subsequently straightens itself; and it can, as shown in the drawing, securely clasp a support. I have seen an object as soft as a young vine-leaf caught by one.

The lower and naked part of the sub-peduncle (C) is likewise

Fig. 10.

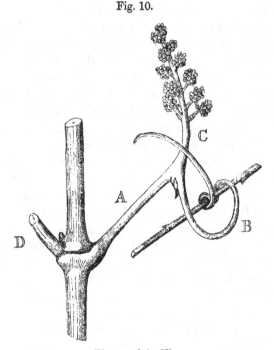

Flower of the Vine.

A. Common Peduncle.
B. Flower-tendril, with a scale at its base.
C. Sub-Peduncle.
D. Petiole of opposite leaf.

slightly sensitive to a rub, and I have seen it distinctly bent round a stick and even partly round a leaf with which it had come into contact. That the sub-peduncle has the same nature as the corresponding branch of the ordinary tendril is well shown when it bears only a few flowers; for in this case it becomes less branched, increases in length, and gains both in sensitiveness and in the power of spontaneous movement. I have twice seen sub-peduncles (C), bearing only from thirty to forty flower-buds, which had be-

come considerably elongated and had completely wound round
sticks, exactly like true tendrils. The whole length of another sub-
peduncle bearing only eleven flower-buds quickly became curved
when slightly rubbed; but even this scanty number of flowers
rendered the stalk less sensitive than the other branch, that is, the
flower-tendril; for the latter after a lighter rub became curved in
a greater degree and more quickly than the sub-peduncle with its
few flowers. I have seen a sub-peduncle thickly covered with
flower-buds, but with one of the higher lateral branchlets bearing
from some cause only two buds, and this one branchlet had become
much elongated and had spontaneously caught hold of an adjoining twig; in fact, it formed a little tendril. The increase of
length in the sub-peduncle (C) with the decreasing number of
its flower-buds is a good instance of the law of compensation.
Hence it is that the whole ordinary tendril is longer than the whole
flower-peduncle; thus, on one and the same plant, the longest
flower-peduncle (measured from the base of the common peduncle
to the tip of the flower-tendril) was $8\frac{1}{2}$ inches in length, whilst the
longest tendril was nearly double this length, namely 16 inches.

The gradation from the ordinary state of the flower-peduncle,
as represented in the drawing (fig. 10), to that of the true tendril
(fig. 9) is perfect. We have seen that the sub-peduncle (C), whilst
still bearing from thirty to forty flower-buds, may become somewhat elongated and partially assume all the characters of the
corresponding branch of the true tendril. From this state we can
trace every stage till we come to a full-sized common tendril,
bearing on the branch which corresponds with the sub-peduncle
one single flower-bud! Hence there can be no doubt that the
tendril is a modified flower-peduncle.

Another kind of gradation well deserves notice. The flower-
tendril (B, fig. 10) sometimes produces a few flower-buds; I
found thirteen and twenty-two on two flower-tendrils on a vine
growing against my house; in this state they retain their characteristic qualities of sensitiveness and spontaneous movement, but
in a somewhat lessened degree. On vines in hothouses, so many
flowers are occasionally produced by the flower-tendrils that a
double bunch of grapes is the result; and this is technically called
by gardeners a "cluster." In this state the whole bunch of
flowers presents scarcely any resemblance to a tendril; and,
judging from the facts already given, it would probably possess
little power of clasping a support, or of spontaneous movement.
Such flower-peduncles closely resemble in structure those borne

by the next genus, *Cissus*. This genus, as we shall immediately see, produces well-developed tendrils and ordinary bunches of flowers; but there is no gradation between the two states. If the genus *Vitis* were unknown, the boldest believer in the modification of species would never, I suppose, have surmised that the same individual plant, at the same period of growth, would have yielded every possible gradation between ordinary flower-stalks for the support of the flowers and fruit, and tendrils used exclusively for climbing. But the vine clearly gives us this case; and it seems to me as striking and curious an instance of transition as can well be conceived.

Cissus discolor.—The young shoots show no more movement than can be accounted for by daily variations in the action of the light. The tendrils, however, revolve with much regularity, following the sun, and, in the plants observed by me, swept circles of about 5 inches in diameter. Five circles were completed in the following times:—4 h. 45 m., 4 h. 50 m., 4 h. 45 m., 4 h. 30 m., and 5 h. The same tendril continues revolving during three or four days. The tendrils are from 3½ to 5 inches in length; they are formed of a long foot-stalk, bearing two short branches, which in old plants again bifurcate. The two branches are not of quite equal length; and, as with the vine, the longer one has a scale at its base. The tendril stands vertically upwards; the extremity of the shoot is bent abruptly downwards; and this position is probably of service in keeping it out of the way of the revolving tendril.

The two branches whilst young are highly sensitive; for I found a touch with a pencil so gentle as only just to move the tendril which was borne at the end of a long flexible shoot, sufficed to cause it to become perceptibly curved in four or five minutes; the tendril became straight again in rather above one hour. A loop of soft thread weighing one-seventh of a grain was thrice tried, and caused the tendrils to become curved in 30 or 40 m.: half this weight produced no effect. The long foot-stalk is much less sensitive, for slight rubbing produced no effect; but prolonged contact with a stick caused it to bend. The two terminal branches are sensitive on all sides; if a number of tendrils be just touched on different sides, two branches of the one on their inner sides, two on their outer sides, or both branches on the same side, in about a quarter of an hour they present a curiously different appearance. If a branch be touched at the same time with equal force on opposite sides, both sides are equally stimulated and there is no movement. At the beginning of my work, and before examining this plant, I had observed only those tendrils which are sensitive on one side, and these when lightly pressed between the finger and thumb become curved; but on thus pinching many times the tendrils of this *Cissus* no curvature ensued, and I was at first falsely led to infer that they were not at all sensitive to a touch.

Cissus antarcticus.—The tendrils on a young plant were thick and straight, with the tips a little curved; when the concave surface was rubbed with some force they very slowly became curved, and subsequently became straight again. Hence they are much less sensitive than the tendrils of the last species; but they made two revolutions, following the sun, rather more rapidly, viz. in 3 h. 30 m. and 4 h. The internodes do not revolve.

Ampelopsis hederacea, or *Virginian Creeper.*—In this plant also the internodes do not move more than apparently can be accounted for by the varying action of the light. The tendrils are from 4 to 5 inches in length; the main stem sends off several lateral branches, which have their tips curved, as may be seen in fig. 11, A. They exhibit no true spontaneous revolving movement, but turn, as was long ago observed by Andrew Knight*, from the light to the dark. I have seen several tendrils move through an angle of 180° to the dark side of a case in less than 24 hours; but the movement is sometimes very much slower. The several lateral branches often move independently of each other, and sometimes irregularly, without any apparent cause. These tendrils are less sensitive to a touch than any others observed by me: by gentle but repeated rubbings with a twig, the lateral branches, but not the main stem, became in the course of three or four hours slightly curved; but they seemed to have hardly any power of again straightening themselves. The tendrils of a plant which crawled over a large box-tree clasped several of the branches. But I have repeatedly seen the tendrils come into contact with sticks, and then withdraw from them. When they meet with a flat surface of wood, or a wall (and this is evidently what they are adapted for), they turn all their branches towards it, and, spreading them widely apart, bring their hooked tips laterally into contact with it. In effecting this, the several branches, after touching the surface, often rise up, place themselves in a new position, and again come down into contact with it.

In the course of about two days after a tendril has arranged its branches so as to press on any surface, the curved tips swell, become bright red, and form on their under-sides the well-known

* Trans. Phil. Soc. 1812, p. 314.

little disks or cushions, which adhere firmly to the surface. In one case these tips became slightly swollen in 38 h. after coming into contact with a brick; in another case they were considerably swollen in 48 h., and in an additional 24 h. they were firmly attached to a smooth board; and lastly, the tips of a younger tendril not only swelled but became attached to a stuccoed wall in 42 h. These adhesive disks resemble, except in colour and in being larger, those of *Bignonia capreolata*. When they were developed in contact with a ball of tow, fibres were separately enveloped, but not in so effective a manner as with *B. capreolata*. Disks are never developed, as far as I have seen, without the stimulus of at least temporary contact with some object. They are generally first formed on one side of the curved tip, the whole of which often becomes so much changed, that a line of green unaltered tissue can be traced only along the concave surface. When, however, a tendril has clasped a cylindrical stick, an irregular rim or disk is formed along the inner surface at some little distance from the curved tip; this was also observed (S. 71) by Mohl. The disks consist of enlarged cells, with smooth projecting hemispherical surfaces, coloured red, and at first gorged with fluid (see section given by Mohl, S. 70), but they ultimately become woody.

As the disks can almost immediately adhere firmly to such smooth surfaces as planed and painted wood, or to the polished leaf of the ivy, this alone would render it probable that some cement is secreted, as has been asserted to be the case (quoted by Mohl, S. 71) by Malpighi. I removed a number of disks formed during the previous year from a stuccoed wall, and placed them in warm water, diluted acetic acid and alcohol during many hours; but the attached grains of silex were not loosened: immersion in sulphuric ether for 24 h. much loosened them; but warmed essential oils (I tried oil of thyme and peppermint) in the course of a few hours completely released every atom of stone. This seems to prove that some resinous cement is secreted; the quantity secreted, however, must be small; for when a plant ascended a thinly whitewashed wall, the disks adhered firmly to the whitewash; but as the cement never penetrated the thin layer, they were easily withdrawn, together with little scales of the whitewash. It must not be supposed that the attachment is by any means exclusively effected by the cement; for the cellular outgrowth completely envelopes every minute and irregular projection, and insinuates itself into every crevice.

A tendril which has not become attached to any body, does not contract spirally; and in course of a week or two shrinks into the finest thread, withers and drops off. An attached tendril, on the other hand, contracts spirally, and thus becomes highly elastic; so that when the main foot-stalk is pulled, the strain is equally distributed to all the attached disks. For a few days after the

Fig. 11.

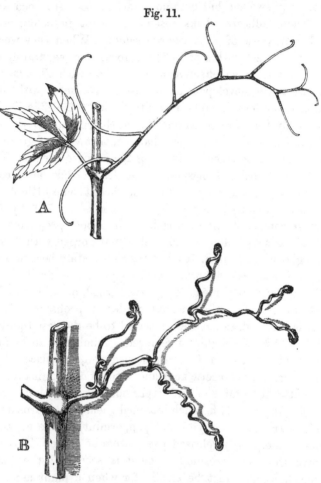

Ampelopsis hederacea

A. Tendril, with the young leaf.

B. Tendril, several weeks after its attachment to a wall, with the branches thickened and spirally contracted, and with the extremities developed into disks. The unattached branches have withered and dropped off.

attachment of the disks, the tendril remains weak and brittle, but it rapidly increases in thickness and acquires great strength: during the following winter it ceases to live, but remains firmly attached to the stem and to the surface of attachment. In the accompanying diagram we may compare the differences of a tendril (B) some weeks after attachment to a wall, with one (A) from the same plant, fully grown but unattached. That the change in the nature of the tissues of the tendril, as well as the act of spiral contraction, is consequent on the formation of the disks, is well shown by any lateral branches which have not become attached; for these in a week or two wither and drop off, in the same manner as does a whole tendril when unattached. The gain in strength and durability in a tendril after its attachment is something wonderful. There are tendrils now adhering to my house which are still strong and have been exposed to the weather in a dead state for fourteen or fifteen years. One single lateral branchlet of a tendril, estimated to be at least ten years old, was still elastic and supported a weight of exactly two pounds. This tendril had five disk-bearing branches of equal thickness and of apparently equal strength; so that this one tendril, after having been exposed during ten years to the weather, would have resisted a strain of ten pounds!

SAPINDACEÆ.—*Cardiospermum halicacabum.*—In this family, as in the last, the tendrils are modified flower-peduncles. In our present plant there are no organs exclusively used for climbing like ordinary tendrils; but the two lateral branches of the main flower-peduncle have been converted into a pair of tendrils, corresponding with the single "flower-tendril" of the common vine. The main peduncle is thin, stiff, and from 3 to 4½ inches in length. Near the summit, above two little bracts, it divides into three branches. The middle one divides and redivides, and bears the flowers; ultimately it grows half as long again as the two other modified branches. These latter are the tendrils; they are at first thicker and longer than the middle branch, but never become more than an inch in length. They taper to a point and are flattened, with the lower clasping surface destitute of hairs. At first they project straight up; but soon diverging, they spontaneously curl downwards so as to become symmetrically and elegantly hooked, as represented in the diagram. They are now, whilst the flower-buds are still small, ready for action.

Fig. 12.

C. halicacabum.

Upper part of the flower-peduncle with its two tendrils.

The two or three upper young internodes steadily revolve; those on one plant made two circles, against the course of the sun, in 3 h. 12 m.; in a second plant the same course was followed, and the two were completed in 3 h. 41 m.; in a third plant the internodes followed the sun, and made two circles in 3 h. 47 m. The average rate of these six revolutions was 1 h. 46 m. The stem shows no tendency to twine spirally round a support; but the allied tendril-bearing genus *Paullinia* is said (Mohl, S. 4) to be a twiner. By the revolving movement, the flower-peduncles, which stand up above the end of the shoot, are carried round and round; but when the internodes were securely tied, the long and thin peduncles themselves were seen to be in continued and sometimes rapid movement from side to side. They swept a wide space, but only occasionally moved in a moderately regular elliptical course. By these combined movements one of the two short hooked tendrils, sooner or later, catches hold of some twig or branch, and then it curls round and securely grasps it. These tendrils are, however, but slightly sensitive; for by rubbing their under surfaces only a slight movement was slowly produced. I hooked a tendril on to a twig; and in 1 h. 45 m. it had curved considerably inwards; in 2 h. 30 m. it formed a ring; and in from 5 to 6 hours from being first hooked, it closely grasped the stick. A second tendril acted at nearly the same rate; but I observed one that took 24 hours before it curled twice round a thin twig. Tendrils which have caught nothing spontaneously curl, after the interval of several days, closely up into a helix. Those which have curled round some object soon become a little thicker and tougher. The long and thin main peduncle, though spontaneously moving, is not sensitive and never clasps a support. It never contracts spirally. Such contraction would apparently have been of service to the plant in climbing; nevertheless it climbs pretty well without this aid. The seed-capsules, though light, are of enormous size (hence its English name of Balloon-vine), and as two or three are carried on the same peduncle, the tendrils arising close to them may possibly be of service in preventing these balloons from being dashed to pieces by the wind. In the hothouse they served simply for climbing.

The position of the tendrils alone suffices to show their homological nature; but in two instances one of the tendrils produced at its tip a flower; this, however, did not prevent the tendril acting properly and curling round a twig. In a third case the two lateral branches which ought to have existed as tendrils, both produced flowers like the central branch, and had quite lost their tendril-structure.

I have only seen, but was not enabled carefully to observe, one other climbing Sapindaceous plant, namely *Paullinia*. It was not in flower, yet thus it bore fine long forked tendrils, differing from *Cardiospermum*. So that, in its tendrils, *Paullinia* apparently bears the same relation to *Cardiospermum* that *Cissus* does to *Vitis*.

PASSIFLORACEÆ.—After reading the discussion and facts given by Mohl (S. 47) on the nature of the tendrils in this family, no one can doubt that they are modified flower-peduncles. The tendrils and true flower-peduncles rise close side by side; and my son, Mr. W. E. Darwin, made sketches for me of their earliest state of development in the hybrid *P. floribunda*. The two organs at first appear as a single papilla which gradually divides; so that I presume the tendril is a modified branch of a single flower-peduncle. My son found one very young tendril surmounted by traces of floral organs, exactly like those on the summit of the true flower-peduncle at the same early age.

Passiflora gracilis.—This well-named, elegant, annual species differs from the other members of the group, observed by me, in the young internodes having the power of revolving. It exceeds all other climbing plants in the rapidity of its movements, and all tendril-bearers in the sensitiveness of its tendrils. The internode which carries the upper active tendril and which likewise carries one or two young immature internodes, made three revolutions, following the sun, at an average rate of 1 h. 4 m.; it then made, the day becoming very hot, three other revolutions at an average rate of between 57 and 58 m.; so that the average rate of all six revolutions was 1 h. 1 m. The apex of the tendril described ellipses, sometimes narrow and long, sometimes broad and long, with their longer axes inclined in slightly different directions. The plant can ascend a thin upright stick by the aid of its tendrils; but the stem is too stiff for it to twine spirally round a stick, even when not interfered with by the tendrils, which had been successively pinched off at an early age.

When the stem was secured, the tendrils were seen to revolve in nearly the same manner and at the same rate as the internodes. The tendrils are very thin, delicate, and straight, with the exception of the tips, which are a little curved; they are from 7 to 9 inches in length. A half-grown tendril was not sensitive; but when nearly full-grown they are extremely sensitive. A single delicate touch on the concave surface of the tip soon caused it to curve, and in two minutes it formed an open helix. A loop of soft thread weighing $\frac{1}{32}$nd of a grain (equal to only two millegrammes) placed most gently on the tip, thrice plainly caused it to curve; as twice did a bent bit of thin platina wire weighing $\frac{1}{50}$th of a grain; but this latter weight, when left suspended, did not suffice to cause permanent curvature. These trials were made under a bell-glass, so that the loops of thread and wire were not agitated by the wind. The movement after a touch is very rapid: I took hold of the lower part of several tendrils and then touched with a thin twig their concave tips, and watched them carefully through a lens; the tips plainly began to bend in the following times—31, 25, 32, 31, 28, 39, 31, and 30 seconds; so that the movement was generally perceptible in half a minute after the touch, but once plainly in 25 seconds. One of the tendrils which thus became bent in 31 seconds had been touched two hours previously and had coiled into a helix; thus in this interval it had straightened itself and had perfectly recovered its sensibility.

I repeated the experiment made on the *Echinocystis*, and placed several plants of this *Passiflora* so close together that the tendrils were repeatedly dragged over each other; but no curvature ensued. I likewise repeatedly flirted small drops of water from a brush on many tendrils, and syringed others so violently that the whole tendril was dashed about, but they never became curved. The impact from the drops of water on my hand was felt far more plainly than that from the loops of thread (weighing $\frac{1}{32}$nd of a grain) when allowed to fall on it; and these loops, which caused the tendrils to become curved, had been placed most gently on them. Hence it is clear, either that the tendrils are habituated to the touch of other tendrils and to that of drops of rain, or that they are sensitive only to prolonged though excessively slight pressure. To show the difference in the kind of sensitiveness in different plants and likewise to show the force of the syringe used, I may add that the lightest jet from it instantly caused the leaves of a *Mimosa* to close; whereas the loop of thread weighing $\frac{1}{32}$nd of a grain, when rolled into a ball and gently placed on the glands at the bases of the leaflets of the *Mimosa*, caused no action. Had I space, I could advance much more striking cases in plants both belonging to the same family, of one being excessively sensitive to the lightest pressure if prolonged, but not to a brief impact; and of another plant equally sensitive to impact, but not to slight though prolonged pressure.

Passiflora punctata.—The internodes do not move; but the tendrils regularly revolve. One that was about half-grown and very sensitive made three revolutions, opposed to the course of the sun, in 3 h. 5 m., 2 h. 40 m., and 2 h. 50 m.; perhaps it might have travelled more quickly when nearly full-grown. The plant was placed in front of a window, and I ascertained that, as with twining stems so with these tendrils, the light accelerated the movement in one direction and retarded it in the other; the semicircle towards the light being performed in one instance in 15 m., and in a second instance in 20 m. less time than that required by the semicircle towards the dark end of the room. Considering the extreme tenuity of these tendrils, the action of the light on them is remarkable. The tendrils are long, and, as just stated, very thin, with the tip slightly curved or hooked. The concave side is extremely sensitive to a touch—even a single touch causing it to curl inwards; it subsequently straightens itself, and is again ready to act. A loop of soft thread weighing $\frac{1}{14}$th of a grain caused the extreme tip to bend; at another time I tried to hang the same little loop on an inclined tendril, but three times it slid off; yet this extraordinarily slight degree of friction sufficed to make the tip curl. The tendril, though so sensitive, does not move very quickly after a touch, no conspicuous change being observable until 5 or 10 m. had elapsed. The convex side of the tip is not sensitive to a touch or to a suspended loop of thread. In one instance I observed a tendril revolving with the convex side of the tip forwards, and on coming into contact with a stick it merely scraped up and past the obstacle and was not able to clasp it; whereas tendrils revolving with the concave side of their tips forward promptly seize any object in their path.

Passiflora quadrangularis.—This is a very distinct species. The tendrils are thick, long, and stiff; they are sensitive to a touch only towards the extremity and on the concave surface. When a stick was so placed that the middle of the tendril came into contact with it, no curvature ensued. In the hothouse a tendril made two revolutions each in 2 h. 22 m.; in my cooler study one was completed in 3 h., and a second in 4 h. The internodes do not revolve; nor do those of the hybrid *P. floribunda.*

Tacsonia manicata.—Here again the internodes do not revolve. The tendrils are moderately thin and long; one made a narrow ellipse in 5 h. 20 m., and the next day a broad ellipse in 5 h. 7 m. The extremity being lightly rubbed on the concave surface, became just perceptibly curved in 7 m., clearly curved in 10 m., and hooked in 20 m.

We have seen that the tendrils in the last three families, namely the Vitaceæ, Sapindaceæ, and Passifloraceæ, are modified flower-peduncles. This is likewise the case, according to De Candolle (as quoted by Mohl), with the tendrils of *Brunnichia*, one of the Polygonaceæ. In two or three species of *Modecca*, one of the Papayaceæ, the tendrils, as I hear from Prof. Oliver, occasionally bear flowers and fruit; so that at least they are axial in their nature.

Spiral contraction of Tendrils.—This movement, which shortens the tendrils and renders them elastic, commences in half a day or in a day or two after their extremities have caught some object. There is no such movement in any leaf-climber, with the exception of an occasional trace of it in the petioles of *Tropæolum tricolorum*. On the other hand, it occurs with all tendrils after they have seized some object, with the few following exceptions,—namely *Corydalis claviculata*, but then this plant might still be called a leaf-climber; *Bignonia unguis* and its close allies, and the *Cardiospermum*; though these tendrils are so short that the contraction could hardly take place, and would be quite superfluous; and *Smilax aspera*, the tendrils of which, though rather short, offer a more marked exception. In the *Dicentra*, whilst young, the tendrils are short and do not contract spirally, but only become slightly flexuous; the longer tendrils, however, borne by older plants contract spirally. I have seen no other exceptions to the rule that all tendrils, after clasping by their extremities a support, contract spirally. When, however, the tendril of any plant of which the stem happens to be immoveably fixed, catches some fixed object, it does not contract, simply because it cannot; this, however, rarely occurs. In the common Pea only the lateral branches, and not the central stem of the tendril, contract; and with most plants, such as the Vine, Passiflora, Bryony, the basal portion never contracts into a spire.

I have said that in *Corydalis claviculata* the end of the leaf or the tendril (for this part may be indifferently thus designated) does not contract into a spire. The branchlets, however, of the tendril, after they have wound round thin twigs, become deeply sinuous or zigzag; and this may be the first indication of the process of spiral contraction. Moreover the whole end of the petiole or tendril, if it seizes nothing, ultimately bends abruptly downwards and inwards, showing that its inferior surface contracts; and this may be confidently looked at as the first indication of the power of spiral contraction. For with all true tendrils when they contract spirally, it is the lower surface, as Mohl (S. 52) has remarked, which contracts. If the inferior surface of the extremity of a free tendril were to contract quite regularly, it would roll itself up into a flat helix, as occurs with the *Cardiospermum*; but if it were to contract in the least on one side, or if the basal portion were first to contract (as does occur), the long free extremity could not be rolled up within the basal part, or if the tip were held during the contraction, as when a tendril has caught some object,—in all these cases the inevitable result would be the formation not of a helix, but of a spire, such as free and caught tendrils form in the act of contraction.

Tendrils of many kinds of plants, if they catch nothing, contract after an interval of several days or weeks into a close spire; but in these cases the movement takes place after the tendril has lost its revolving power and has partly or wholly lost its sensibility, and hangs downwards; this, as we shall presently see, is a quite useless movement. The spiral contraction of unattached tendrils is a much slower process than that of attached tendrils: young tendrils which have caught a support and are spirally contracted may be constantly seen on the same stem with much older tendrils, unattached and uncontracted. In the *Echinocystis* I have seen a tendril with the two lateral branches clasped to twigs and contracted into beautiful spires, whilst the main branch which had caught nothing remained for many days afterwards uncontracted. In this plant I once observed a main branch after it had caught a stick become spirally flexuous in 7 h., and spirally contracted in 18 h. Generally the tendrils of the *Echinocystis* begin to contract in from 12 h. to 24 h. after catching something; whilst its unattached tendrils do not begin to contract until two or three or even more days have elapsed after the revolving movement has ceased. I will give one other case: a full-grown tendril of *Passiflora quadrangularis* which had caught a stick began in 8 h. to contract, and in 24 h. several spires were formed; a younger tendril, only two-thirds grown, showed the first trace of contraction in two days after clasping a stick, and in two additional days had formed several spires; hence, apparently, the contraction does not begin in a tendril until it is grown to nearly its full length. Another young tendril of about the same age and length as the last did not catch any object; it acquired its full length in four days; in six additional days it first became flexuous, and in two more days had formed one complete spire. This first spire was formed towards the basal end of the tendril, and the contraction steadily but slowly progressed towards the apex; but the whole was not closely wound up until 21 days had elapsed from the first observation, that is until 17 days after the tendril was fully grown.

The best proof of the intimate connexion between the spiral contraction of a tendril and the previous act of clasping a support, is afforded by those tendrils which, when caught, invariably contract into a spire, whilst as long as they remain unattached they continue straight, though dependent, and thus wither and drop off. The tendrils of *Bignonia*, which are modified leaves, thus behave, as do the tendrils of the three genera of Vitaceæ, and these are modified flower-peduncles. The tendrils, however, of *Eccremocarpus*, which is allied to *Bignonia*, contract spirally even when they have caught nothing. The uncaught tendrils of the *Cardiospermum*, and to a certain extent those of the *Mutisia*, roll themselves up not into a spire, but into a helix.

The spiral contraction which ensues after a tendril has caught a support is of high service to all tendril-bearing plants; hence its almost universal occurrence with plants of widely different orders. When a shoot is inclined and its tendril has caught an object above, the spiral contraction drags up the shoot. When the shoot is upright, the growth of the internodes, subsequently to the tendrils having seized some object above, would slacken the stem were it not for the spiral contraction, which draws up the internodes as they increase in length. Thus there is no waste of growth, and the stretched stem ascends by the shortest course. We have seen in the *Cobæa*, when a terminal branchlet of the tendril has caught a stick, how well the spiral contraction of its branches successively brings them one after the other into contact with the stick, until the whole tendril has grasped it in an inextricable knot. When a tendril has caught a yielding object, this is sometimes enveloped and still further secured by the spiral folds, as I have seen with *Passiflora quadrangularis*; but this action is of little importance.

A far more important service rendered by the spiral contraction is that the tendrils are thus made highly elastic. As was previously remarked under *Ampelopsis*, the strain is thus equally distributed to the several attached branches of a branched tendril; and this must render the whole tendril far stronger, as branch after branch cannot separately break. It is this elasticity which saves both branched and simple tendrils from being torn away during stormy weather. I have more than once gone on purpose during a gale to watch a Bryony growing in an exposed hedge, with its tendrils attached to the surrounding bushes; and as the thick or thin branches were tossed to and fro by the wind, the attached tendrils, had they not been excessively elastic, would instantly have been torn off and the plant thrown prostrate. But as it was, the Bryony safely rode out the gale, like a ship with two anchors down, and with a long range of cable ahead to serve as a spring as she surges to the storm.

With respect to the exciting cause of the spiral contraction, little can be said. After reading Prof. Oliver's interesting paper * on the hygroscopic contraction of legumes, I allowed a number of different kinds of tendrils to dry slowly, but no spiral contraction ensued; nor did this occur with the tendrils of the Bryony when placed in water, diluted alcohol, and syrup of sugar. We know that the act of clasping a support leads to a change in the nature of their tissues; and we call this a vital action, and so we must call the spiral contraction. The contraction is not related to the spontaneous revolving power, for it occurs in tendrils, such as those of *Lathyrus grandiflorus* and *Ampelopsis hederacea*, which do not revolve. It is not necessarily related to the curling of the tips round a support, as we see in the case of the *Ampelopsis* and *Bignonia capreolata*, in which the development of the adherent disks suffices to induce the contraction. Yet it certainly seems to stand in some close relation to the curling or clasping movement due to contact with a support; for not only does it soon follow this act, but the spiral contraction generally begins close to the curled extremity, and travels down towards the base, as if the whole tendril tried to imitate the movement of its extremity. If, however, a tendril be very slack, the whole length seems to become almost simultaneously at first flexuous and then spiral. The spiral contraction of a tendril when unattached cannot serve any of the useful ends just described; it does not occur with many kinds of tendrils which contract when attached; and when it does occur, it supervenes, as we have seen, only after a considerable interval of time. It may almost be likened to certain instinctive or habitual movements performed by animals under circumstances rendering them manifestly useless.

When an uncaught tendril contracts spirally, the spire always runs in the same direction from tip to base. A tendril, on the other hand, which has caught a support by its extremity, invariably becomes twisted in one part in one direction, and in another part in the opposite direction; the oppositely turned spires being

* Trans. Linn. Soc. vol. xxiv. 1864, p. 415.

separated by short straight portions. This curious and symmetrical structure has been noticed by several botanists, but has not been

Fig. 13.

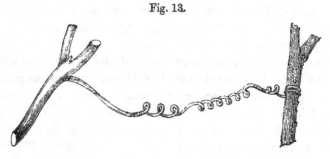

A caught tendril of *Bryonia dioica*, spirally contracted in reversed directions.

explained*. It occurs without exception with all tendrils which after catching any object contract spirally, but is of course most conspicuous in the longer tendrils; it never occurs with uncaught tendrils; and when this appears to have happened, it will be found that the tendril had originally seized some object and had afterwards been torn free. Commonly all the spires at one end of a caught tendril run in one direction, and all those at the other end in the opposite direction, with a single short straight portion in the middle; but I have seen a tendril with the spires alternately turning five times in opposite directions, with straight portions between them; and M. Léon has seen seven or eight such alternations. Whether the spires turn several times in opposite directions, or only once, there are as many turns in the one direction as in the other. For instance, I gathered ten long and short caught tendrils of the Bryony, the longest with 33, and the shortest with only 8 spiral turns; and the number of turns in one direction was in every case the same (within one) as in the opposite direction.

The explanation of this curious little fact is not difficult; I will not attempt any geometrical reasoning, but will give only practical illustrations. In doing this, I shall first have to allude to a point which was almost passed over when treating of Twining-plants. If we hold in our left hand a bundle of parallel strings, we can with our right hand turn these round and round, and imitate the revolving movement of a twining plant, and the strings do not become twisted. But if we now at the same time hold a stick in our left hand, in such a position that the strings become

* See M. Isid. Léon in Bull. Soc. Bot. de France, tom. v. 1858, p. 680.

spirally turned round it, they will inevitably become twisted. Hence a straight coloured line, painted along the internodes of a twining plant before it has wound round a support, becomes twisted or spiral after it has so wound round. I painted a red line on the straight internodes of a *Humulus*, *Mikania*, *Ceropegia*, *Convolvulus*, and *Phaseolus*, and saw it become twisted as the plant wound round a stick. It is possible that the stems of some plants by spontaneously turning on their own axes, at the proper rate and in the proper direction, might avoid becoming twisted; but I have seen no such case.

In the above illustration, the parallel strings were wound round a stick; but this is by no means necessary, for if wound into a hollow coil (as can be done with a narrow slip of elastic paper) there is the same inevitable twisting of the axis. Hence when a tendril, which is free at its end, coils itself into a spire, it must either become twisted along its whole length (and this is a case which I have never seen), or the free extremity must turn round as many times as there are spires formed. It was hardly neces- sary to observe this fact; but I did so by affixing little paper vanes to the extreme points of the tendrils of the *Echinocystis* and *Passiflora quadrangularis*; and as the tendril contracted itself into successive spires, the vane slowly revolved.

We can now understand the meaning of the spires being in- variably turned in opposite directions in those tendrils which, having caught some object, are thus fixed at both ends. Let us suppose a caught tendril to make thirty spiral turns in one direc- tion; the inevitable result will be that it will become thirty times twisted on its own axis. This twisting not only would require considerable force, but, as I know by trial, would burst the ten- dril before the thirty turns were completed. Such a case never really occurs; for, as already stated, when a tendril has caught a support and has spirally contracted, there are always as many turns in one direction as in the other; so that the twisting of the axis in the one direction is exactly compensated by that in the other. We can further see how the tendency is given to make coils in an opposite direction to those, whether turned to the right or to the left, which are first made. Take a piece of string, and let it hang down with the lower end fixed to the floor; then wind the upper end (holding the string quite loosely) spirally round a per- pendicular pencil, and this will twist the lower part of the string; after it has been sufficiently twisted, it will be seen to curve itself into an open spire, with the curves running in an opposite direc-

H

tion to those round the pencil, and consequently with a straight
piece of string between the opposite spires. In short, we have
given to the string the regular spiral arrangement of a tendril
caught at both ends. The spiral contraction generally begins at
the extremity which has clasped a support; and these first-formed
spires give a twist to the axis of the tendril, which necessarily
inclines the basal part into an opposite spiral curvature. I cannot resist giving one other illustration, though superfluous: when
a haberdasher winds up ribbon for a customer, he does not wind
it into a single coil; for, if he did, the ribbon would twist itself as
many times as there were coils; but he winds it into a figure of
eight on his thumb and little finger, so that he alternately takes
turns in opposite directions, and thus the ribbon is not twisted.
So it is with tendrils, with this sole difference, that they take
several consecutive turns in one direction and then the same
number in an opposite direction; but in both cases the self-twist-
ing is equally avoided.

Summary on the Nature and Action of Tendrils.—In the concluding remarks I shall have to allude to some points which may
be here passed over. In the majority of tendril-bearing genera
the young internodes revolve in more or less broad ellipses, like
those made by twining plants; but the figures described, when
carefully traced, generally form irregular ellipsoidal spires. The
rate of revolution in different plants varies from one to five hours,
and consequently in some cases is more rapid than with any
twining plant, and is never so slow as with those many twiners,
which take more than five hours for each revolution. The direction is variable even in the same individual plant. In *Passiflora*,
the internodes of only one of the species have the power of revolving. The Vine is the weakest revolver observed by me, apparently exhibiting only a trace of a former power. In the *Eccremocarpus* the movement is interrupted by many long pauses. Some,
but very few, tendril-bearing plants can spirally twine up an upright stick. Although the twining-power has generally been lost
by tendril-bearers, either from the stiffness or shortness of the
internodes, from the size of the leaves, or from other unknown
causes, the revolving movement well serves to bring the tendrils
into contact with surrounding objects.

The tendrils also have the power of revolving in the same manner
and generally at the same rate with the internodes. The movement begins whilst the tendril is young, but is at first slow. In
Bignonia littoralis even the mature tendrils moved much slower

than the internodes. In all cases the conditions of life must be favourable for the perfect action of the tendrils. Generally both internodes and tendrils revolve together; in other cases, as in *Cissus*, *Cobæa*, and most Passifloræ, the tendrils alone revolve; in other cases, as with *Lathyrus aphaca*, the internodes alone move, carrying with them the motionless tendrils; and, lastly (and this is the fourth possible case), neither internodes nor tendrils spontaneously revolve, as with *Lathyrus grandiflorus* and the *Ampelopsis*. In most Bignonias, in the *Eccremocarpus*, *Mutisia*, and the Fumariaceæ, the petioles as well as the tendrils, together with the internodes, all spontaneously move together.

The tendrils revolve by the curvature of their whole length, excepting the extremity and excepting the base, which parts do not move, or move but little. The movement is of the same nature as that of the revolving internodes. Hence, if a line be painted along that surface which at the time happens to be convex, the line becomes first lateral and then concave, and ultimately again convex. This experiment can be tried only on the thicker tendrils, which are not affected by a thin crust of dried paint. The extremities, however, of the tendrils, which so often are slightly curved or hooked, never reverse their curvature; and in this respect they differ from the extremities of the shoots of twining plants, which not only reverse their curvature, or at least become periodically straight, but curve in a greater degree than the lower portions. But, in fact, the tendril answers to the upper internode of the several revolving internodes of a twining plant; and in the former part of this paper it was explained how the several internodes move together by the whole successively curving to all points of the compass. There is, however, in many cases this unimportant difference, that the curving tendril is separated from the curving internode by a rigid petiole. There is also another difference, namely, that the summit of the shoot, which in itself has no power of revolving, projects above the point from which the tendril arises; but the summit of the shoot is generally thrown on one side, so as to be out of the way of the revolutions swept by the tendril. In those plants in which the terminal shoot is not sufficiently out of the way, the tendril, as we have seen with the *Echinocystis*, as soon as it comes in its revolving course to this point, stiffens and straightens itself, and, rising up vertically, passes over the obstacle.

All tendrils are sensitive, but in very various degrees, to contact with any object, and curve towards the touched side. With several plants a single touch, so slight as only just to move the highly flexible tendril, is enough to induce curvature. *Passiflora gracilis* has the most sensitive tendrils which I have seen: a bit of platina wire $\frac{1}{50}$th of a grain in weight, gently placed on the concave point, caused two tendrils to become hooked, as did (and this perhaps is a better proof of sensitiveness) a loop of soft, thin cotton thread weighing $\frac{1}{32}$nd of a grain, or about two milli-grammes. With the tendrils of several other plants, loops weighing $\frac{1}{16}$th of a grain sufficed. The point of the tendril of the *Passiflora gracilis* distinctly began to move in 25 seconds after a touch. Asa Gray saw movement in the tendrils of the Cucurbi-taceous genus, *Sicyos*, in 30 seconds. The tendrils of some other plants, when lightly rubbed, move in a few minutes; in the *Dicentra* in half-an-hour; in the *Smilax* in an hour and a quarter or a half; and in the *Ampelopsis* still more slowly. The curling movement consequent on a single touch continues to increase for a considerable time, then ceases; after a few hours the tendril uncurls itself, and is again ready for action. When very light weights were suspended on tendrils of several plants and caused them to curve, these seemed to become accustomed to so slight a stimulus, and straightened themselves, as if the loops had been removed. It makes no difference, as far as I have seen, what sort of object a tendril touches, with the remarkable exception of drops of water in the case of the extremely sensitive tendrils of *Passiflora gracilis* and of the *Echinocystis*; hence we are led to infer that they have become habituated to showers of rain. As I made no observations with this view on other tendrils, I cannot say whether there are more cases of this adaptation. Moreover adjoining tendrils rarely catch each other, as we have seen with the *Echinocystis* and *Passiflora*, though I have seen this occur with the Bryony.

Tendrils of which the extremities are slightly curved or bowed are sensitive only on the concave surface; other tendrils, such as those of the *Cobæa* (though furnished with minute horny hooks) and those of *Cissus discolor*, are sensitive on all sides. Hence the tendril of this latter plant, when stimulated by a touch of equal force on opposite sides, does not bend. In the tendril of the *Mutisia* the inferior and lateral surfaces are sensitive, but not the upper surface. With branched tendrils, the several branches all act alike; but in the *Hanburya* the lateral spur-like branch does not acquire (for a reason which has been explained) its sensitiveness nearly so soon as the main branch. The lower or basal part of many tendrils is either not at all sensitive or sensitive only to prolonged contact. Hence we see that the sensitiveness of tendrils is a special and localized capacity, quite independent of the power of spontaneously moving; for the curling of the terminal portion from a touch does not in the least interrupt the spontaneous revolving movement of the lower part. In *Bignonia unguis* and its close allies the petioles of the leaves, as well as the tendrils, are sensitive to a touch.

Twining plants when they come into contact with a stick, curl round it invariably in the direction of their revolving movement; but tendrils curl indifferently to either side, in accordance with the position of the stick and the side which is first touched. The clasping-movement of the extremity apparently is not steady, but vermicular in its nature, as may be inferred from the manner in which the tendrils of the *Echinocystis* slowly crawled round a smooth stick.

As with a few exceptions tendrils spontaneously revolve, it may be asked, Why are they endowed with sensitiveness?—why, when they come into contact with a stick, do they not, like a twining plant, spirally wind round it? One reason may be that in most cases they are so flexible and thin that, when brought into contact with a stick, they would yield, and their revolving movement would not be arrested; they would thus be dragged onwards and away from the stick. Moreover the sensitive extremities have no revolving power, and could not by this means curl round any object. With twining plants, on the other hand, the extremity of the shoot spontaneously bends more than any other part; and this is of high importance to the ascending power of the plant, as may be seen on a windy day. It is, however, possible that the slow movement of the basal and stiffer parts of certain tendrils, which wind round sticks placed in their course, may be analogous to that of twining plants. I doubt this; but I hardly attended sufficiently to this point, and it would be difficult to distinguish between a movement due to extremely dull sensitiveness and that resulting from the arrestment of the lower part together with the continued movement of the terminal part of a tendril.

Tendrils which are only three-fourths grown, and perhaps even when younger, but not whilst extremely young, have the power of revolving and of grasping any object which they may touch. These two capacities generally commence at about the same period, and fail when the tendril is full grown. But in the *Cobæa* and *Passiflora punctata* the tendrils began revolving in a quite useless manner, before they became sensitive. In the *Echinocystis* they retained their sensitiveness for some time after they had ceased revolving and had drooped downwards ; in this position, even if they should seize any object, they could be of little or no use in supporting the stem.   It is a rare circumstance thus to be able to detect any imperfection or superfluity in tendrils—organs which are so admirably adapted for the functions which they have to perform ; but we see that they are not always absolutely perfect, and it would be rash to assume that any existing tendril has reached the utmost limit of perfection.

Some tendrils have their revolving motion accelerated and retarded in moving to and from the light ; others, as with the Pea, seem indifferent to its action ; others move from the light to the dark, and this aids them in an important manner in finding a support.   In *Bignonia capreolata* the tendrils bend from the light to the dark, like a banner from the wind.   In the *Cobæa* and *Eccremocarpus* the extremities alone twist and turn about, so as to bring their finer branches and hooks into close contact with any surface, or into dark crevices and holes.   This latter movement is one of the best adapted exhibited by tendrils.

A short time after a tendril (with some rare exceptions) has caught a support, it contracts spirally ; but the manner of contraction and the several important advantages thus gained have been so lately discussed, that nothing need be here said on the subject.   Again, tendrils soon after catching a support grow much stronger and thicker, and sometimes in a wonderful degree durable ; and all this shows how much their internal tissues must change.   Tendrils which have caught nothing soon shrink and wither ; in some species of *Bignonia* they disarticulate and fall off like leaves in autumn.

Any one who did not closely study tendrils of various kinds would probably infer that their action would always be uniform. This is the case with most kinds of tendrils, of which the extremities simply curl round objects of any moderate degree of thickness, and of various shapes or natures.   But *Bignonia* shows us what diversity of action there may be in the tendrils of even closely allied species.   In all the nine species of this genus observed by me the young internodes revolved vigorously ; as did the petioles of nearly all, but in very unequal degrees ; in three of the species the petioles were sensitive to contact ; the tendrils of all are sensitive to contact, and likewise revolve, but in some of the species in a very feeble manner.   In the first-described unnamed species, the tendrils, in shape like a bird's foot, are of no service when the stem spirally ascends a thin upright stick, but they can seize any twig or branch lying beneath them ; but when the stem spirally ascends a somewhat thicker stick, a slight degree of sensitiveness in the petioles is brought into play, and they wind their tendrils round the stick. In *B. unguis* and *B. Tweedyana* the sensitiveness, as well as the power of movement, in the petioles is greatly augmented ; and the tendrils and petioles are thus inextricably wound together round thin upright sticks ; but the stem, in consequence, does not twine so well : *B. Tweedyana*, in addition, emits aërial roots which adhere to the stick. In *B. venusta* the tendrils have lost the bird's-foot structure, and are converted into long three-pronged grapnels ; these exhibit a conspicuous power of spontaneous movement; the petioles, however, have lost their sensitiveness. The stem can spirally twine round an upright stick, and is aided in its ascent by the tendrils alternately seizing the stick some way above and then spirally contracting. In this and all the following species the tendrils spirally contract after seizing any object. In *B. littoralis* and *B. Chamberlaynii* the tendrils, which have the same structure as in *B. venusta*, and the non-sensitive petioles and the internodes all spontaneously revolve. The stem, however, cannot spirally twine, but ascends an upright stick by both tendrils, seizing it above. In *B. littoralis* the tips of the tendrils become developed into adhesive disks. In *B. speciosa* and *B. picta* we have similar powers of movement, but the plant cannot spirally twine round a stick; it can, however, ascend by clasping it with one or both of its unbranched tendrils, on their own level ; and these exhibit the strange, apparently useless, habit of continually inserting their pointed ends into minute crevices and holes. In *B. capreolata* the stem twines in an imperfect manner; the much-branched tendrils revolve in a capricious manner, and they have the power of bending in a conspicuous manner from the light to the dark; their hooked extremities, even whilst immature, crawl into any crevice, or, when mature, seize any thin projecting point ; in both cases they develope adhesive disks, which have the power of enveloping by growth the finest fibres.

In the allied *Eccremocarpus* the internodes, petioles, and tendrils all spontaneously revolve together ; its much-branched tendrils resemble those of *Bignonia capreolata*, but they do not turn from the light ; and their bluntly hooked extremities, which arrange themselves so neatly to any surface, do not form adhesive disks ; they act best when each seizes a few thin stems, like the culms of a grass, which they afterwards draw together by their spiral con-

traction into a firm bundle. In the *Cobæa* the tendrils alone
revolve; these are divided into many fine branches, terminating
in sharp little hooks, which crawl into crevices, and are turned by
an excellently adapted movement to any object that is seized. In
the *Ampelopsis*, on the other hand, there is little or no power of
revolving in any part: the branched tendrils are but little sensitive to contact; their hooked extremities cannot seize any thin
object; they will not even clasp a stick, unless in extreme need
of a support; but they turn from the light to the dark, and,
spreading out their branches in contact with any nearly flat surface, the disks are developed. These can adhere, by the secretion
of some cement, to a wall, or even to a polished surface; and this
is more than the disks of the *Bignonia capreolata* can effect.

The formation and rapid growth of these adherent disks is one
of the most remarkable peculiarities in the structure and functions
of tendrils. We have seen that such disks are formed by two
species of *Bignonia*, by the *Ampelopsis*, and, according to Naudin[*],
by the Cucurbitaceous genus *Peponopsis adhærens*. Their development, apparently in all cases, depends on the stimulus from
contact. It is not a little singular that three families so widely
distinct as the Bignoniaceæ, Vitaceæ, and Cucurbitaceæ should all
have species bearing tendrils with this same remarkable pecu-
liarity. Most tendrils, after they have clasped any object, rapidly
increase in strength and thickness throughout their whole length;
but some tendrils, when wound round a support either by the
middle or the extremity, become swollen at these points in a
remarkable manner; thus I have seen the clasped portion of a
tendril of the *Bignonia Chamberlaynii* grown twice as thick as the
free basal portion, and become wonderfully rigid. In the *Anguria* the lower surface of the tendril, after it has wound round a
stick, forms a coarsely cellular layer, which closely fits the wood,
but is not adherent; in the *Hanburya* a similar layer is developed,
which is adherent; lastly, in the *Peponopsis* adherent disks are
formed at the tips of the tendrils. These three last-named genera
belong to the Cucurbitaceæ, so that, in this one family, we have a
nearly perfect gradation from a common tendril to one that forms
an adherent disk at its tip; the one small step which is wanted is
a tendril in a state between that of the *Anguria* and *Hanburya*—
that is, adherent only in a slight degree or occasionally.

Finally, it may be added that America, which so abounds with
arboreal animals, as has lately been insisted on by Mr. Bates,

* Annales des Sc. Nat. Bot. 4th series, tom. xii. p. 89.

likewise, according to Mohl and Palm, abounds with climbing plants; and, of the tendril-bearing plants examined by me, the most admirably constructed come from this grand continent, namely, the several species of *Bignonia, Eccremocarpus, Cobæa,* and *Ampelopsis.*

Part IV.—HOOK-CLIMBERS.—ROOT-CLIMBERS.—CONCLUDING REMARKS.

Hook-climbers.—In my introductory remarks, I stated that, besides the great class of twining plants, with the subordinate divisions of leaf-climbers and tendril-bearers, there were hook- and root-climbers. I mention the former only to say that with the few which I have examined, namely, *Galium aparine, Rubus australis,* and some climbing Roses, there is no spontaneous revolving movement. If indeed they possessed this power, and were capable of twining, such plants would be placed in the previous great class: thus the Hop, which is a twiner, has reflexed hooks as large as those of the *Galium*; some other twiners have stiff reflexed hairs; *Dipladenia* has a circle of blunt spines at the base of its leaves; one tendril-bearing plant alone, as far as I have seen, namely, *Smilax aspera,* is furnished with spines. Some few plants, which apparently depend solely on their hooks, are excellent climbers, as certain Palms in the New and Old Worlds. Even some of the climbing Roses will ascend the walls of a tall house, if covered with a trellis: how this is effected I know not; for the young shoots of one such Rose, when placed in a pot in a window, bent irregularly towards the light during the day and from it during the night, like any other plant; so that it is not easy to understand how the shoots can get under a trellis close to a wall.

Root-climbers.—A good many plants come under this class, and are excellent climbers. One of the most remarkable is the *Marcgravia umbellata,* which in the tropical forests of South America, as I hear from Mr. Spruce, grows in a curiously flattened manner against the trunks of trees, here and there putting forth claspers (roots), which adhere to the trunk, and, if the latter be slender, completely embrace it. When this plant has climbed to the light, it sends out free and rounded branches, clad with sharp-pointed leaves, wonderfully different in appearance from those borne by the stem, as long as it is adherent. This surprising difference in the leaves I have observed in a plant of *M. dubia* in my hothouse Root-climbers, as far as I have seen, namely, the Ivy (*Hedera helix*), *Ficus repens*, and *F. barbatus*, have no power of movement, not even from the light to the dark. As previously stated, the *Hoya carnosa* (Asclepiadaceæ) is a spiral twiner, and can likewise adhere by rootlets even to a flat wall; the tendril-bearing *Bignonia Tweedyana* emits roots, which curve half round and adhere to thin sticks. The *Tecoma radicans* (Bignoniaceæ), which is closely allied to many spontaneously revolving species, climbs by rootlets; but its young shoots apparently move about rather more than can be accounted for by the varying action of the light.

I have not closely observed many root-climbers, but can give one curious little fact. *Ficus repens* climbs up walls just like Ivy; when the young rootlets were made to press lightly on slips of glass, they emitted (and I observed this several times), after about a week's interval, minute drops of clear fluid, not in the least milky like that exuded from a wound. This fluid was slightly viscid, but could not be drawn out into threads; it had the remarkable property of not drying. One drop, about the size of half a pin's head, I slightly spread out, and scattered on it some minute grains of sand. The slip of glass was left exposed in a drawer during hot and dry weather, and, if the fluid had been water, it would certainly have dried in one or two minutes; but it remained fluid, closely surrounding each grain of sand, during 128 days: how much longer it would have remained I cannot say. Some other rootlets were left in contact with the glass for about ten days or a fortnight, and the drops of fluid secreted by them were rather larger, and so viscid that they could be drawn out into threads. Some other rootlets were left in contact during twenty-three days, and these were firmly cemented to the glass. Hence we may conclude that the rootlets first secrete a slightly viscid fluid, and that they subsequently absorb (for we have seen that it will not dry by itself) the watery parts, and ultimately leave a cement. When the rootlets were torn from the glass, atoms of yellowish matter were left on it, which were partly dissolved by a drop of bisulphide of carbon; and this extremely volatile fluid was rendered, by what it had dissolved, very much less volatile.

As the bisulphide of carbon has so strong a power of softening indurated caoutchouc*, I soaked in it during a short time many

---

* Mr. Spiller has recently shown (Chemical Society, Feb. 16, 1865), in a paper on the oxidation of india-rubber, that this substance, when exposed to the air in a fine state of division, gradually becomes converted into brittle, resinous matter, very similar to shell-lac.

rootlets of a plant which had grown up a plaistered wall. Attached
to two sets of rootlets on the same branch, I found very many
extremely thin threads of a transparent, not viscid, excessively
elastic substance, precisely like caoutchouc. These threads, at
one end, proceeded from the bark of the rootlet, and at the other
end were firmly attached to transparent particles of silex and other
hard substances. There could be no mistake in this observation,
for I played with the threads for a long time, under the microscope,
drawing them out with the dissecting-needles and letting them
spring back again. Yet, as I looked repeatedly at other rootlets,
similarly treated, and could never discover these elastic threads,
I infer that the branch had probably been slightly moved from
the wall at some critical period, whilst the fluid secreted from the
rootlets was in the act of drying and of changing its nature
through the absorption of its watery parts. The genus *Ficus*
abounds with caoutchouc, and from the facts here given we may
infer that this substance, at first in solution and ultimately modified into an unelastic cement, is used by *Ficus repens* to cement
its rootlets to any object which it may ascend. Whether most
other plants, which climb by their rootlets, emit any cement I do
not know; but the rootlets of the Ivy, placed against glass, barely
adhered to it, yet secreted a little yellowish matter. I may add,
that the rootlets of *Marcgravia dubia* can adhere firmly to smooth
painted wood.

Vanilla aromatica emits aërial roots a foot in length, which
point straight down to the ground. According to Mohl (S. 49),
these crawl into crevices, and, when they meet with a thin support, wind round it, like tendrils. A plant which I kept was
young, and did not form long roots; but on placing thin sticks in
contact with them, they certainly bent, in the course of about a day,
a little to that side, and adhered by their rootlets to the wood;
but they did not bend quite round the sticks, and afterwards they
repursued their downward course. If these rootlets are really
sensitive to contact and bend to the touched side, in this case
the class of root-climbers blends into that of tendril-bearers.
According to Mohl, the rootlets of certain species of *Lycopodium*
likewise act as tendrils.

Concluding Remarks.

Plants become climbers, in order, it may be presumed, to reach
the light, and to expose a large surface of leaves to its action
and to that of the free air. This is effected by climbers with

wonderfully little expenditure of organized matter, in comparison with trees, which have to support a load of heavy branches by a massive trunk.   Hence, no doubt, it arises that there are in all quarters of the world so many climbing plants belonging to so many different orders.   These plants have been here classed under three heads :—Firstly, hook-climbers, which are, at least in our temperate countries, the least efficient of all, and can climb only in the midst of an entangled mass of vegetation.   Secondly, root-climbers, which are excellently adapted to ascend naked faces of rock : when they climb trees, they are compelled to keep much in the shade ; they cannot pass from branch to branch, and thus cover the whole summit of a tree, for their rootlets can adhere only by long-continued and close contact with a steady surface.   Thirdly, the great class of spiral-twiners, with the subordinate divisions of leaf-climbers and tendril-bearers, which together far exceed in number and in perfection of mechanism the climbers of the two previous classes.   These plants, by their power of spontaneously revolving and of grasping objects with which they come in contact, can easily pass from branch to branch, and securely ramble over a wide and sun-lit surface.

I have ranked twiners, leaf- and tendril-climbers as subdivisions of one class, because they graduate into each other, and because nearly all have the same remarkable power of spontaneously re-volving.   Does this gradation, it may be asked, indicate that plants belonging to one subdivision have passed, during the lapse of ages, or can pass, from one state to the other ; has, for instance, a tendril-bearing plant assumed its present structure without having previously existed as either a leaf-climber or a twiner ?   If we consider leaf-climbers alone, the idea that they were primordially twiners is forcibly suggested.   The internodes of all, without ex-ception, revolve in exactly the same manner as twiners ; and some few can still twine well, and many others in a more or less imper-fect manner.   Several leaf-climbing genera are closely allied to other genera which are simple twiners.   It should be observed, that the possession by a plant of leaves with their petioles or tips sensitive, and with the consequent power of clasping any object, would be of very little use, unless associated with revolving inter-nodes, by which the leaves could be brought into contact with surrounding objects.   On the other hand, revolving internodes, without other aid, suffice to give the power of climbing ; so that, unless we suppose that leaf-climbers simultaneously acquired both capacities, it seems probable that they were at first twiners, and subsequently became capable of grasping a support, which, as we shall presently see, is a great additional advantage.

From analogous reasons, it is probable that tendril-bearing plants were primordially twiners, that is, are the descendants of plants having this power and habit. For the internodes of the majority revolve, like those of twining plants; and, in a very few, the flexible stem still retains the capacity of spirally twining round an upright stick. With some the internodes have lost even the revolving power. Tendril-bearers have undergone much more modification than leaf-climbers; hence it is not surprising that their supposed primordial revolving and twining habits have been lost or modified more frequently than with leaf-climbers. The three great tendril-bearing families in which this loss has occurred in the most marked manner are the Cucurbitaceæ, Passifloraceæ, and Vitaceæ. In the first the internodes revolve; but I have heard of no twining form, with the exception (according to Palm, S. 29. 52) of *Momordica balsamina*, and this is only an imperfect twiner. In the other two families I can hear of no twiners; and the internodes rarely have the power of revolving, this power being confined to the tendrils; nevertheless the internodes of *Passiflora gracilis* have this power in a perfect manner, and those of the common Vine in an imperfect degree: so that at least a trace of the supposed primordial habit is always retained by some members of the larger tendril-bearing groups.

On the view here given, it may be asked, Why have nearly all the plants in so many aboriginally twining groups been converted into leaf-climbers or tendril-bearers? Of what advantage could this have been to them? Why did they not remain simple twiners? We can see several reasons. It might be an advantage to a plant to acquire a thicker stem, with short internodes bearing many or large leaves; and such stems are ill fitted for twining. Any one who will look during windy weather at twining plants will see that they are easily blown from their support; not so with tendril-bearers or leaf-climbers, for they quickly and firmly grasp their support by a much more efficient kind of movement. In those plants which still twine, but at the same time possess tendrils or sensitive petioles, as some species of *Bignonia, Clematis*, and *Tropæolum*, we can readily observe how incomparably more securely they grasp an upright stick than do simple twiners. From possessing the power of movement on contact, tendrils can be made very long and thin; so that little organic matter is expended in their development, and yet a wide circle is swept.

Tendril-bearers can, from their first growth, ascend along the outer branches of any neighbouring bush, and thus always keep in the full light; twiners, on the contrary, are best fitted to ascend bare stems, and generally have to start in the shade. In dense tropical forests, with crowded and bare stems, twining plants would probably succeed better than most kinds of tendril-bearers; but the majority of twiners, at least in our temperate regions, from the nature of their revolving movement, cannot ascend a thick trunk, whereas this can be effected by tendril-bearers, if the trunks carry many branches or twigs; and in some cases they can ascend by special means a trunk without branches, but with rugged bark.

The object of all climbing plants is to reach the light and free air with as little expenditure of organic matter as possible; now, with spirally ascending plants, the stem is much longer than is absolutely necessary; for instance, I measured the stem of a kidney-bean, which had ascended exactly two feet in height, and it was three feet in length : the stem of a pea, ascending by its tendrils, would, on the other hand, have been but little longer than the height gained. That this saving of stem is really an advantage to climbing plants I infer from observing that those that still twine, but are aided by clasping petioles or tendrils, generally make more open spires than those made by simple twiners. Moreover, such plants very generally, as was observed over and over again with the several leaf-climbers, after taking one or two turns in one direction, ascend for a space straight, and then reverse the direction of their spire. By this means they ascend to a considerably greater height, with the same length of stem, than would otherwise be possible; and they can do it with safety, as they secure themselves at intervals by their clasping petioles.

We have seen that tendrils consist of various organs in a modified state, namely, leaves and flower-peduncles, and perhaps branches and stipules. The position alone generally suffices to show when a tendril has been formed from a leaf; and in *Bignonia* the lower leaves are often perfect, whilst the upper ones terminate in a tendril in place of a terminal leaflet; in *Eccremocarpus* I have seen a lateral branch of a tendril replaced by a perfect leaflet; and in *Vicia sativa*, on the other hand, leaflets are sometimes replaced by tendril-branches; and many other such cases could be given. But he who believes in the slow modification of species will not be content simply to ascertain the homological nature of different tendrils; he will wish to learn, as far as possible, by what steps parts acting as leaves or as flower-peduncles can have wholly changed their function, and have come to serve as prehensile organs.

In the whole group of leaf-climbers abundant evidence has been given that an organ, still subserving its proper function as a leaf, may become sensitive to a touch, and thus grasp an adjoining object. In several leaf-climbers true leaves spontaneously revolve; and their petioles, after clasping a support, grow thicker and stronger. We thus see that true leaves may acquire all the leading and characteristic qualities of tendrils, namely, sensitiveness, spontaneous movement, and subsequent thickening and induration. If their blades or laminæ were to abort, they would form true tendrils. And of this process of abortion we have seen every stage; for in an ordinary tendril, as in that of the Pea, we can discover no trace of its primordial nature; in *Mutisia clematis*, the tendril, in shape and colour, closely resembles a petiole with the denuded midribs of its leaflets; and occasionally vestiges of laminæ are retained or reappear. Lastly, in four genera in the same family of the Fumariaceæ we see the whole gradation; for the terminal leaflets of the leaf-climbing *Fumaria officinalis* are not smaller than the other leaflets; those of the leaf-climbing *Adlumia cirrhosa* are greatly reduced; those of the *Corydalis claviculata* (a plant which may indifferently be called a leaf-climber or tendril-bearer) are either reduced to microscopical dimensions or have their blades quite aborted, so that this plant is in an actual state of transition; and, finally, in the *Dicentra* the tendrils are perfectly characterized. Hence, if we were to see at the same time all the progenitors of the *Dicentra*, we should almost certainly behold a series like that now exhibited by the above-named four genera. In *Tropæolum tricolorum* we have another kind of passage; for the leaves which are first formed on the young plant are entirely destitute of laminæ, and must be called tendrils, whilst the later-formed leaves have well-developed laminæ. In all cases, in the several kinds of leaf-climbers and of tendril-bearers, the acquirement of sensitiveness by the mid-ribs of the leaves apparently stands in the closest relation with the abortion of their laminæ or blades.

On the view here given, leaf-climbers were primordially twiners, and tendril-bearers (of the modified leaf division) were primordially leaf-climbers. Hence leaf-climbers are intermediate in nature between twiners and tendril-bearers, and ought to be related to both. This is the case: thus the several leaf-climbing species of the *Antirrhineæ*, of *Solanum*, of *Cocculus*, of *Gloriosa* are related to other genera in the same family, or even to other species in the same genus, which are true twiners. On the other hand, the leaf-climbing species of *Clematis* are very closely allied to the tendril-bearing *Naravelia* : the Fumariaceæ include closely allied genera which are leaf-climbers and tendril-bearers. Lastly, one species of *Bignonia* is both a leaf-climber and a tendril-bearer, and other closely allied species are twiners.

Tendrils of the second great division consist of modified flower-peduncles. In this case likewise we have many interesting transitional states. The common Vine (not to mention the *Cardiospermum*) gives us every possible grade from finely developed tendrils to a bunch of flower-buds, bearing the single usual lateral flower-tendril. And when the latter itself bears some flowers, as we know is not rarely the case, and yet retains the power of clasping a support, we see the primordial state of all those tendrils which have been formed by the modification of flower-peduncles.

According to Mohl and others, some tendrils consist of modified branches : I have seen no such case, and therefore of course know nothing of any transitional states, if such occur. But *Lophospermum* at least shows us that such a transition is possible ; for its branches spontaneously revolve, and are sensitive to contact. Hence, if the leaves of some of the branches were to abort, they would be converted into true tendrils. Nor is it so improbable as it may at first appear that certain branches alone should become modified, the others remaining unaltered ; for we have seen with certain varieties of *Phaseolus* that some of the branches are thin and flexible and twine, whilst other branches on the same plant are stiff and have no such power.

If we inquire how the petiole of a leaf, or the peduncle of a flower, or a branch first becomes sensitive and acquires the power of bending towards the touched side, we get no certain answer. Nevertheless an observation by Hofmeister[*] well deserves attention, namely, that the shoots and leaves of all plants, whilst young, move after being shaken ; and it is almost invariably young petioles and young tendrils, whether formed of modified leaves or flower-peduncles, which move on being touched ; so that it would appear as if these plants had utilized and perfected a widely distributed and incipient capacity, which capacity, as far as we can see, is of no service to ordinary plants. If we

---

* Quoted by F. Cohn, in his remarkable memoir, "Contractile Gewebe im Pflanzenreiche," Abhand. der Schlesichen Gesell. 1861, Heft i. S. 35.

further inquire how the stems, petioles, tendrils, and flower-peduncles of climbing plants first acquired their power of spontaneously revolving, or, to speak more accurately, of successively bending to all points of the compass, we are again silenced, or at most can only remark, that the power of movement, both spontaneous and from various stimuli, is far more common with plants, as we shall presently see, than is generally supposed to be the case by those who have not attended to the subject. There is, however, the one remarkable case of the *Maurandia semperflorens*, in which the young flower-peduncles spontaneously revolve in very small circles, and bend themselves, when gently rubbed, to the touched side; yet this plant certainly profits in no way by these two feebly developed powers. A rigorous examination of other young plants would probably show some slight spontaneous movements in the peduncles and petioles, as well as that sensitiveness to shaking observed by Hofmeister. We see at least in the *Maurandia* a plant which might, by a little augmentation of qualities which it already possesses, come first to grasp a support by its flower-peduncles (as with *Vitis* or *Cardiospermum*) and then, by the abortion of some of its flowers, acquire perfect tendrils.

There is one interesting point which deserves notice. We have seen that some tendrils have originated from modified leaves, and others from modified flower-peduncles; so that some are foliar and others axial in their homological nature. Hence it might have been expected that they would have presented some difference in function. This is not the case. On the contrary, they present the most perfect identity in their several remarkable characteristics. Tendrils of both kinds spontaneously revolve at about the same rate. Both, when touched, bend quickly to the touched side, and afterwards recover themselves and are able to act again. In both the sensitiveness is either confined to one side or extends all round the tendril. They are either attracted or repelled by the light. The latter case is seen in the foliar tendrils of *Bignonia capreolata* and in the axial tendrils of the *Ampelopsis*, both of which move from the light. The tips of the tendrils in these two plants become, after contact, enlarged into disks, which are at first adhesive by the secretion of some cement. Tendrils of both kinds, soon after grasping a support, contract spirally; they then increase greatly in thickness and strength. When we add to these several points of identity the fact of the petiole of the *Solanum jasminoides* assuming the most characteristic feature of the axis, namely, a closed ring of woody vessels, we can hardly avoid asking, whether the difference between foliar and axial organs can be of so fundamental a nature as is generally supposed to be the case*.

We have attempted to trace some of the stages in the genesis of climbing plants. But, during the endless fluctuations in the conditions of life to which all organic beings have been exposed, it might have been expected that some climbing plants would have lost the habit of climbing. In the cases given of certain South African plants belonging to great twining families, which in certain districts of their native country never twine, but reassume this habit when cultivated in England, we have a case in point. In the leaf-climbing *Clematis flammula*, and in the tendril-bearing Vine, we see no loss in the power of climbing, but only a remnant of that revolving-power which is indispensable to all twiners, and is so common, as well as so advantageous, to most climbers. In *Tecoma radicans*, one of the Bignoniaceæ, we see a last and doubtful trace of the revolving-power.

With respect to the abortion of tendrils, certain cultivated varieties of *Cucurbita pepo* have, according to Naudin†, either quite lost these organs or bear semi-monstrous representatives of them. In my limited experience, I have met with only one instance of their natural suppression, namely, in the common Bean. All the other species of *Vicia*, I believe, bear tendrils; but the Bean is stiff enough to support its own stem, and in this species, at the end of the petiole where a tendril ought to have arisen, a small pointed filament is always present, about a third of an inch in length, and which must be considered as the rudiment of a tendril. This may be the more safely inferred, because I have seen in young unhealthy specimens of true tendril-bearing plants similar rudiments. In the Bean these filaments are variable in shape, as is so frequently the case with all rudimentary organs, being either cylindrical, or foliaceous, or deeply furrowed on the upper surface. It is a rather curious little fact, that many of these filaments when foliaceous have dark-coloured glands on their lower surfaces, like those on the stipules, which secrete a sweet fluid; so that these rudiments have been feebly utilized.

One other analogous case, though hypothetical, is worth giving. Nearly all the species of *Lathyrus* possess tendrils; but *L. nissolia* is destitute of them. This plant has leaves, which must have

---

* Mr. Herbert Spencer has recently argued ('Principles of Biology,' 1865, p. 37 *et seq.*) with much force that there is no fundamental distinction between foliar and axial organs in plants.

† Annales des Sc. Nat. 4th series, Bot. tom. vi. 1856, p. 31.

struck every one who has noticed them with surprise, for they are quite unlike those of all common papilionaceous plants, and resemble those of a grass. In *L. aphaca* the tendril, which is not highly developed (for it is unbranched, and has no spontaneous revolving-power), replaces the leaves, the latter in function being replaced by the large stipules. Now if we suppose the tendrils of *L. aphaca* to become flattened and foliaceous, like the little rudimentary tendrils of the Bean, and the large stipules, not being any longer wanted, to become at the same time reduced in size, we should have the exact counterpart of *L. nissolia*, and its curious leaves are at once rendered intelligible to us.

It may be added, as it will serve to sum up the foregoing views on the origin of tendril-bearing plants, that if these views be correct, *L. nissolia* must be descended from a primordial spirally-twining plant; that this became a leaf-climber; that first part of the leaf and then the whole leaf became converted into a tendril, with the stipules by compensation greatly increased in size*; that this tendril lost its branches and became simple, then lost its revolving-power (in which state it would resemble the tendril of the existing *L. aphaca*), and afterwards losing its prehensile power and becoming foliaceous would no longer be called a tendril. In this last stage (that of the existing *L. nissolia*) the former tendril would reassume its original function of a leaf, and its lately largely developed stipules, being no longer wanted, would decrease in size. If it be true that species become modified in the course of ages, we may conclude that *L. nissolia* is the result of a long series of changes, in some degree like those just traced.

The most interesting point in the natural history of climbing plants is their diverse powers of movement; and this led me on to their study. The most different organs—the stem, flower-peduncle, petiole, mid-ribs of the leaf or leaflets, and apparently aërial roots—all possess this power.

In the first place, the tendrils place themselves in the proper position for action, standing, for instance in the *Cobœa*, vertically upwards, with their branches divergent and their hooks turned outwards, and with the young terminal shoot thrown on one side; or, as in *Clematis*, the young leaves temporarily curve themselves downwards, so as to serve as grapnels.

---

* Moquin-Tandon (Eléments de Tératologie, 1841, p. 156) gives the case of a monstrous Bean, in which a case of compensation of this nature was suddenly effected; for the leaves had completely disappeared and the stipules had grown to an enormous size.

Secondly, if the young shoot of a twining plant, or if a tendril, be placed in an inclined position, it soon bends upwards, though completely secluded from the light. The guiding stimulus to this movement is no doubt the attraction of gravity, as Andrew Knight showed to be the case with germinating plants. If a succulent shoot of almost any plant be placed in an inclined position in a glass of water in the dark, the extremity will, in a few hours, bend upwards; and if the position of the shoot be then reversed, the now downward-bent shoot will reverse its curvature; but if the stolon of a Strawberry, which has no tendency to grow upwards, be thus treated, it will curve downwards in the direction of, instead of in opposition to, the force of gravity. As with the Strawberry, so it is generally with the twining shoots of the *Hibbertia dentata,* which climbs laterally from bush to bush; for these shoots, when bent downwards, show little and sometimes no tendency to curve upwards.

Thirdly, climbing plants, like other plants, bend towards the light by a movement closely analogous to that incurvation which causes them to revolve. This similarity in the nature of the movement was well seen when climbing plants were kept in a room, and their first movements in the morning towards the light, and their subsequent revolving movements, were traced on a bell-glass. We have also seen that the movement of a revolving shoot, and in some cases of a tendril, is retarded or accelerated in travelling from or to the light. In a few instances tendrils bend in a conspicuous manner towards the dark. Many authors speak as if the movement of a plant towards the light was as directly the result of the evaporation or of the oxygenation of the sap in the stem, as the elongation of a bar of iron from an increase in its temperature. But, seeing that tendrils are either attracted to or repelled by the light, it is more probable that their movements are only guided and stimulated by its action, in the same manner as they are guided by the force of attraction from or towards the centre of gravity.

Fourthly, we have in stems, petioles, flower-peduncles, and tendrils the spontaneous revolving movement which depends on no outward stimulus, but is contingent on the youth of the part and on its vigorous health, which again of course depends on proper temperature and the other conditions of life. This is perhaps the most interesting of all the movements of climbing plants, because it is continuous. Very many other plants exhibit spontaneous movements, but they generally occur only once during the life of the plant, as in the movements of the stamens and pistils, &c., or at intervals of time, as in the so-called sleep of plants.

Fifthly, we have in the tendrils, whatever their homological nature may be, in the petioles and tips of the leaves of leaf-climbers, in the stem in one case, and apparently in the aërial roots of the *Vanilla*, movements—often rapid movements—from contact with any body. Extremely slight pressure suffices to cause the movement. These several organs, after bending from a touch, become straight again, and again bend when touched.

Sixthly, and lastly, most tendrils, soon after clasping a support, but not after a mere temporary curvature, contract spirally. The stimulus from the act of clasping some object seems to travel slowly down the whole length of the tendril. Many tendrils, moreover, ultimately contract spontaneously even if they have caught no object; but this latter useless movement occurs only after a considerable lapse of time.

We have seen how diversified are the movements of climbing plants. These plants are numerous enough to form a conspicuous feature in the vegetable kingdom; every one has heard that this is the case in tropical forests; but even in the thickets of our temperate regions the number of kinds and of individual plants is considerable, as will be found by counting them. They belong to many and widely different orders. To gain some crude idea of their distribution in the vegetable series, I marked, from the lists given by Mohl and Palm (adding a few myself, and a competent botanist, no doubt, could add many more), all those families in ' Lindley's Vegetable Kingdom ' which include plants in any of our several subdivisions of twiners, leaf-climbers, and tendril-bearers; and these (at least, some in each group) all have the power of spontaneously revolving. Lindley divides Phanerogamic plants into fifty-nine Alliances; of these, no less than above half, namely thirty-five, include climbing plants according to the above definition, hook- and root-climbers being excluded. To these a few Cryptogamic plants must be added which climb by revolving. When we reflect on this wide serial distribution of plants having this power, and when we know that in some of the largest, well-defined orders, such as the Compositæ, Rubiaceæ, Scrophulari-aceæ, Liliaceæ, &c., two or three genera alone, out of the host of genera in each, have this power, the conclusion is forced on our minds that the capacity of acquiring the revolving-power on which most climbers depend is inherent, though undeveloped, in almost every plant in the vegetable kingdom.

It has often been vaguely asserted that plants are distinguished from animals by not having the power of movement. It should rather be said that plants acquire and display this power only when it is of some advantage to them; but that this is of comparatively rare occurrence, as they are affixed to the ground, and food is brought to them by the wind and rain. We see how high in the scale of organization a plant may rise, when we look at one of the more perfect tendril-bearers. It first places its tendrils ready for action, as a polypus places its tentacula. If the tendril be displaced, it is acted on by the force of gravity and rights itself. It is acted on by the light, and bends towards or from it, or disregards it, whichever may be most advantageous. During several days the tendril or internodes, or both, spontaneously revolve with a steady motion. The tendril strikes some object, and quickly curls round and firmly grasps it. In the course of some hours it contracts into a spire, dragging up the stem, and forming an excellent spring. All movements now cease. By growth the tissues soon become wonderfully strong and durable. The tendril has done its work, and done it in an admirable manner.

Printed in the United States
By Bookmasters